aus der Reihe:

Innovationen mit Mikrowellen und Licht

Forschungsberichte aus dem Ferdinand-Braun-Institut, Leibniz-Institut für Höchstfrequenztechnik

Band 49

Carlo Frevert

Optimization of broad-area GaAs diode lasers for high powers and high efficiencies in the temperature range 200-220 K

Herausgeber: Prof. Dr. Günther Tränkle, Prof. Dr.-Ing. Wolfgang Heinrich

Ferdinand-Braun-Institut
Leibniz-Institut
für Höchstfrequenztechnik (FBH)
Gustav-Kirchhoff-Straße 4
12489 Berlin

Tel.   +49.30.6392-2600
Fax   +49.30.6392-2602

E-Mail fbh@fbh-berlin.de
Web   www.fbh-berlin.de

Innovations with Microwaves and Light

Research Reports from the Ferdinand-Braun-Institut, Leibniz-Institut für Höchstfrequenztechnik

Preface of the Editors

Research-based ideas, developments, and concepts are the basis of scientific progress and competitiveness, expanding human knowledge and being expressed technologically as inventions. The resulting innovative products and services eventually find their way into public life.

Accordingly, the *"Research Reports from the Ferdinand-Braun-Institut, Leibniz-Institut für Höchstfrequenztechnik"* series compiles the institute's latest research and developments. We would like to make our results broadly accessible and to stimulate further discussions, not least to enable as many of our developments as possible to enhance everyday life.

Broad-area GaAs-based diode lasers with highest powers and efficiencies are essential components for pumping solid-state lasers in modern high-energy class laser facilities. Operating the diode lasers at lower temperatures promises a significant performance improvement. This work explores ways to best exploit low temperatures for higher powers and efficiencies by optimizing their vertical epitaxial structure. The kW-class power and > 70% efficiency levels achieved in high fill factor bars employing the found designs could justify the deployment of cooled bars in large laser facilities.

We wish you an informative and inspiring reading

Prof. Dr. Günther Tränkle
Director

Prof. Dr.-Ing. Wolfgang Heinrich
Deputy Director

The Ferdinand-Braun-Institut

The Ferdinand-Braun-Institut researches electronic and optical components, modules and systems based on compound semiconductors. These devices are key enablers that address the needs of today's society in fields like communications, energy, health and mobility. Specifically, FBH develops light sources from the visible to the ultra-violet spectral range: high-power diode lasers with excellent beam quality, UV light sources and hybrid laser systems. Applications range from medical technology, high-precision metrology and sensors to optical communications in space. In the field of microwaves, FBH develops high-efficiency multi-functional power amplifiers and millimeter wave frontends targeting energy-efficient mobile communications as well as car safety systems. In addition, compact atmospheric microwave plasma sources that operate with economic low-voltage drivers are fabricated for use in a variety of applications, such as the treatment of skin diseases.

The FBH is a competence center for III-V compound semiconductors and has a strong international reputation. FBH competence covers the full range of capabilities, from design to fabrication to device characterization.

In close cooperation with industry, its research results lead to cutting-edge products. The institute also successfully turns innovative product ideas into spin-off companies. Thus, working in strategic partnerships with industry, FBH assures Germany's technological excellence in microwave and optoelectronic research.

Optimization of broad-area GaAs diode lasers for high powers and high efficiencies in the temperature range 200-220 K

vorgelegt von

Dipl. Phys.
Carlo Frevert
geb. in Bielefeld

Von der Fakultät IV - Elektrotechnik und Informatik
der Technischen Universität Berlin
zur Erlangung des akademischen Grades

Doktor der Naturwissenschaften
- Dr. rer. nat. -

genehmigte Dissertation

Promotionsausschuss:

Vorsitzender:	Prof. Dr. W. Heinrich
Erstgutachter:	Prof. Dr. G. Tränkle
Zweitgutachter:	Prof. Dr. S. Sweeney
Drittgutachter:	Prof. Dr. M. Kneissl

Tag der wissenschaftlichen Aussprache: 05. Dezember 2018

Berlin, 2018

Bibliografische Information der Deutschen Nationalbibliothek
Die Deutsche Nationalbibliothek verzeichnet diese Publikation in der Deutschen
Nationalbibliographie; detaillierte bibliographische Daten sind im Internet über
http://dnb.d-nb.de abrufbar.
1. Aufl. - Göttingen: Cuvillier, 2019
    Zugl.: (TU) Berlin, Univ., Diss., 2018

© CUVILLIER VERLAG, Göttingen 2019
    Nonnenstieg 8, 37075 Göttingen
    Telefon: 0551-54724-0
    Telefax: 0551-54724-21
    www.cuvillier.de

    ISBN 978-3-7369-9944-2
    eISBN 978-3-7369-8944-3

Abstract

Gaas-based high-power broad area (BA) diode lasers are the most efficient light source in converting electrical into optical energy. Their high power densities, small footprint and high reliability make them essential tools for industrial and scientific applications. In addition, their adjustable wavelength (by choice of semiconductor material) enables their use as optical pumps for a wide range of fibre, gas or solid-state lasers. Widespread integration of diode pumped solid-state lasers in worldwide increasing numbers of large high-energy class laser facilities is currently hindered by the cost of the diodes (in \$/W) and cheaper flashlamps are often used in spite of their relatively low repetition rates, poor electro-optical efficiencies and short lifetimes. As the price for a diode laser (DL) is driven by the cost of the semiconductor chip, an increase in its output power will make the device more economically appealing. While lasers bars with an output power of a few hundred watt are already commercially available, research is being conducted to obtain kW-class DL bars. A promising approach to achieve a step size increase in output power and power conversion efficiency of DL bars is to decrease their operation temperature. Cryogenic cooling of the solid-state laser media is commonly used and the implementation of optical pumps into the cooling cycle would be feasible. Though studies have shown the benefit of cryogenic operation of BA lasers in single emitters at low powers, no extensive research has been performed to achieve high power, high efficient laser bars optimized for low temperature operation.

This work focuses on the development of AlGaAs-based DL bars optimized for reaching highest powers and efficiencies at low operation temperatures. Specifically, the quasi continuous wave (QCW) pumping of cryogenically cooled Yb:YAG solid-state lasers is targeted, setting requirements on the wavelength ($\lambda = 940\,\text{nm}$), the pulse conditions (pulse length $\tau = 1.2\,\text{ms}$ and frequency $f = 10\,\text{Hz}$) and the lowest DL operating temperature $T_{\text{HS}} \sim 200\,\text{K}$ consistent with economic cooling. High fill-factor bars for QCW operation are to reach high optical performance with optical output powers of $P \geq 1.5\,\text{kW}$ and power conversion efficiencies of $\eta_{\text{E}} \geq 60\%$ at these power levels. Understanding the efficiency-limiting factors and the behavior at lower temperatures is necessary to design these devices. After an initial assessment of existing high power single emitter DLs using established vertical epitaxial designs, optimizations are performed iteratively in three stages: first, vertical epitaxial designs are studied theoretically, adjusted to the targeted operation temperatures and specific laser parameters are extracted. Secondly, resulting vertical designs are processed into low power single emitters and their electro-optical behavior at low currents is experimentally assessed over a wide range of temperatures. The obtained laser parameters characteristic to the vertical design are then used to extrapolate the laser's performance up to the high targeted currents. Finally, vertical designs promising to reach the targeted values for power and efficiency are processed into high power single emitters and bars which are measured up to the highest currents.

In accordance with other published results, the optical output power at lower temperature significantly increases due to an increasing internal differential efficiency, a lower transparency current and reduced power saturation. With the optical behavior improved, the series resistance R_S is identified as the main limiting factor to efficiency at high power levels. Furthermore, the development of R_S with decreasing temperature is found to be contrary to its behavior expected from semiconductor properties. While R_S in bulk layers of semiconductor with compositions used for the epitaxial layers has been shown to decrease with reduced temperatures in the assessed temperature range, fabricated DL exhibit an increasing R_S determined from low current measurements. Using similar vertical structures with different Al-contents in the waveguide, the series resistance is experimentally shown to be a clear function of effective barrier height around the quantum well. This represents the first observation describing the reason behind the increased resistance at low temperature in GaAs based DL, namely degraded carrier transport into the quantum well. First studies to compensate for the deteriorating carrier transport show a reduction in R_S by implementing a light p-doping into the active region. While the resistance in excess of the predicted values from bulk layer estimations is successfully reduced by 45%, no negative impact on the emitted output power is observed.

Low Al-content waveguides prove to significantly reduce R_S, especially due to a lower excess resistance. While room temperature performance is significantly deteriorated due to the low barriers around the quantum well, no degradation to the optical behavior is seen at the low targeted temperatures in low current measurements. QCW bars employing these low R_S vertical structures together with a double quantum well are fabricated with a fill factor of 69% and a cavity length $L = 4\,\mathrm{mm}$. At 203 K, they reach output powers of 2 kW with an efficiency of 55% at these power levels, emitting at $\sim 940\,\mathrm{nm}$. This represents the highest ever reported output power from a 1 cm wide bar. At the targeted 1.5 kW, efficiency remains above 60%. A second design approach to reduce R_S focuses on epitaxial structures using thin, asymmetric waveguides. Combined with low Al-contents in the waveguide, vertical designs are found that reach 77% peak efficiencies on bar level using the same lateral layout and operating conditions as for the 2 kW results. At powers of 1.5 kW, efficiency is still at 62%, thus fulfilling the target specifications with both efficiencies being the highest reported in laser bars suitable for pump applications. On the technological level, the development of low resistance conduction cooled packages is crucial in reaching the high efficiencies of the mounted bar package.

The successful mitigation of efficiency limiting factors results in vertical designs optimized for low temperature operation. The achieved power and efficiency levels in bars using these epitaxial designs make the DL attractive pump sources in future high energy class laser facilities using cryogenically cooled solid-state lasers.

Kurzfassung

GaAs-basierte Breitstreifen Hochleistungslaserdioden sind die effizientesten Lichtquellen zur Umwandlung elektrischer in optische Leistung. Durch ihre hohen Leistungsdichten, ihre kompakte Bauweise und die langen Lebensdauern sind Diodenlaser (DL) als unverzichtbare Komponenten in vielen industriellen wie wissenschaftlichen Anwendungen im Einsatz. Ihre durch die Wahl des Halbleitermaterials aus einem breiten Spektralbereich wählbare Emissionswellenlänge erlaubt es, verschiedene Arten von Faser-, Gas- und Festkörperlasern optisch zu pumpen. Der hohe Preis der DL (in \$/W) verhindert derzeit noch ihren Einsatz als optische Pumpquelle für Festkörperlaser, welche in einer weltweit zunehmenden Anzahl von Hochleistungs-Laserzentren betrieben werden. Stattdessen werden aus ökonomischen Überlegungen Blitzlampen verwendet, welche allerdings geringe Wiederholraten, niedrige Effizienzen und kurze Lebensdauern aufweisen. Da der Preis eines DL-Bauteils hauptsächlich durch den Preis des Halbleiter-Materials bestimmt wird, führt eine Erhöhung der Ausgangsleistung pro Bauteil zu einer Kostensenkung. Zur Zeit sind DL-Barren mit Ausgangsleistungen von mehreren hundert Watt kommerziell erhältlich, und an einer Steigerung der Leistung in den kW-Bereich wird geforscht. Ein aussichtsreicher Ansatz, Leistung und Effizienz von DL signifikant zu steigern, besteht in der Absenkung ihrer Betriebstemperatur. Da die zu pumpenden Festkörperkristalle oft auf kryogene Temperaturen ($\sim 170\,$K) abgekühlt werden, können auch die optischen Pumpen in den Kühlkreislauf eingebaut und bei tiefen Temperaturen betrieben werden. In mehreren Publikationen wurde die Verbesserungen von optischer Leistung und Effizienz von Einzelemitter-DL bei tiefen Temperaturen und kleinen Ausgangsleistungen gezeigt. Jedoch wurden bisher keine Studien zu DL Hochleistungsbarren bei tiefen Temperaturen durchgeführt.

In der vorgelegten Arbeit wurden AlGaAs-basierte DL Barren entwickelt, die höchste Leistungen und Effizienzen bei tiefen Temperaturen erreichen. Die Laser dienen im *quasi-continuos wave* (QCW) Betrieb als optische Pumpen von Tieftemperatur-gekühlten Yb:YAG Festkörper-Lasern. Dadurch ergeben sich Anforderungen an die Wellenlänge ($\lambda = 940\,$nm), den Pulsbetrieb (Pulslänge $\tau = 1.2\,$ms, Frequenz $f = 10\,$Hz) und die niedrigste Betriebstemperatur ($T_{HS} \sim 200\,$K), welche sich ohne große Änderungen am eingesetzten Kühlkreislauf erreichen lässt. QCW-Barren hoher Belegungsdichte sollten optische Leistungen von $P \geq 1.5\,$kW und Effizienzen von $\eta_E \geq 60\%$ bei diesen Leistungsniveaus erzielen. Grundsätzlich musste dazu das elektro-optische Verhalten von DL bei sinkenden Betriebstemperaturen erforscht und speziell die Effizienz-verringernden Faktoren analysiert werden. Beginnend mit einer Bewertung der Leistungsfähigkeit bei verringerten Temperaturen von DL mit etablierten Vertikalstrukturen, wurden anschließend

Verbesserungen in drei Schritten durchgeführt: zuerst wurden Epitaxiedesigns simuliert und für die angestrebten Temperaturen optimiert. Anschließend wurden die resultierenden Strukturen zu Einzelemittern mit Ausgangsleistung von bis zu 15 W prozessiert, deren elektro-optischen Eigenschaften bei verschiedenen Temperaturen untersucht wurden. Mit Hilfe der extrahierten charakteristischen Laserparameter der verschiedenen Vertikaldesigns wurden die Ausgangsleistungen und Effizienzen bei den angestrebten tiefen Temperaturen und hohen Strömen extrapoliert. Letztlich wurden die Epitaxiestrukturen mit den höchsten erwarteten Leistungen und Effizienzen zu Hochleistungs-Einzelemitter und Barren prozessiert, welche bei Temperaturen $T_{HS} \sim 200\,\text{K}$ bis zu hohen Strömen vermessen wurden.

Im Einklang mit publizierten Ergebnissen erhöhte sich die Ausgangsleistung von DL bei tiefen Temperaturen durch eine gesteigerte interne differentielle Effizienz, die geringere Transparenz-stromdichte und eine reduzierte Leistungssättigung. Da sich die optische Ausgangsleistung signifikant verbesserte, wurde die Effizienz bei hohen Leistungen primär durch den elektrischen Serienwiderstand R_S begrenzt. Es wurde gezeigt, dass R_S mit abnehmender Temperatur steigt, was im Widerspruch zu den Halbleitereigenschaften steht. Obwohl in anderen Arbeiten demonstriert wurde, dass R_S in Volumenhalbleitern, welche mit den hier verwendeten Kompositionen und Dotierungen gewachsen wurden, mit sin-kenden Temperaturen bis 200 K abnimmt, nimmt R_S (bestimmt aus Messungen bei kleinen Strömen) in DL zu. Mithilfe ähnlicher Vertikalstrukturen, die sich nur im Al-Gehalt des Wellenleiters und der n-Mantelschicht unterscheiden, wurde die Abhängigkeit des Serienwiderstandes von der effektiven Barrierenhöhe um den Quantentrog nachgewiesen. Mit dem Zusammenhang von Barrienhöhe und R_S wurde zum ersten Mal die Ursache für den gestiegenen Serienwiderstand in GaAs-basierten DL bei tiefen Temperaturen gegeben, und mit dem verschlechterten Ladungsträgertransport in die aktive Zone erklärt. In ersten Untersuchungen zur Verbesserung des Ladungsträgertransports wurde R_S durch eine niedrige p-Dotierung der aktiven Zone verringert. Der parasitäre Zusatzwiderstand, der über dem erwarteten Widerstand liegt, wurde erfolgreich um 45% verringert und kein negativer Einfluss auf die optische Leistung wurde beobachtet.

Der Einsatz niedriger Al-Konzentration im Wellenleiter reduzierte R_S stark, was sich vor allem aus dem niedrigeren parasitären Zusatzwiderstand ergab. Während sich Leistung und Effizienz bei Raumtemperatur durch die niedrigeren Barrierenhöhen um den Quantentrog stark verringerten, wurde keine Verschlechterung der optischen Leistung bei den angestrebten tiefen Temperaturen beobachtet. QCW-Barren, basierend auf diesen Vertikalstrukturen mit niedrigem R_S, welche auch einen Doppelquantentrog enthielten, wurden mit hohen Belegungsdichten von 69% und $L = 4\,\text{mm}$ langen Resonatoren hergestellt. Bei einer Betriebstemperatur von 203 K erreichten sie Ausgangsleistungen von 2 kW mit einer Effizienz von 55% und einer Emissionswellenlänge von 940 nm. Dies ist die höchste je veröffentlichte Ausgangsleistung eines 1 cm breiten DL-Barrens. Die angestrebten Leistungen von 1.5 kW wurden mit einer Effizienz von über 60% erreicht.

Ein zweiter Design-Ansatz, um Vertikalstrukturen mit niedrigen R_S zu entwickeln, befasste sich mit Epitaxiestrukturen mit dünnen, asymmetrischen Wellenleitern. Die asymmetrischen Vertikaldesigns mit einem niedrigen Al-Gehalt im Wellenleiter erreichten 77% Spitzeneffizienz aus einem Barren der gleichen lateralen, zuvor beschriebenen Eigenschaften bei gleichen Betriebsbedingungen. Bei einer Leistung von 1.5 kW wurde eine

Effizienz von 62% gemessen. Somit wurden die angestrebten Leistungs- und Effizienz-spezifikationen erfüllt und beide Effizienzwerte stellen die höchsten publizierten Effizienzen von QCW-Pumplaserbarren dar. Aus technologischer Sicht war die Entwicklung von passiv gekühlten Aufbauten mit geringem Serienwiderstand unerlässlich zum Erreichen der hohen Effizienzen der aufgebauten DL-Barren.

Der Einfluss Effizienz-verringernder Faktoren wurden erfolgreich reduziert und Vertikaldesigns für niedrige Betriebstemperaturen entwickelt. Die erreichten Leistungs- und Effizienzwerte von DL-Barren mit diesen Epitaxiestrukturen erhöhen die Vorteile von DL als optische Pumpquellen für Tieftemperatur-Festkörperlaser in Hochleistungs-Laserzentren.

List of Publications

The following contributions have been published within the context of this thesis.

Articles

P. Crump, G. Erbert, H. Wenzel, C. Frevert, C.M. Schultz, K.-H. Hasler, R. Staske, B. Sumpf, A. Maaßdorf, F. Bugge, S. Knigge, and G. Tränkle, *Efficient High-Power Laser Diodes*, IEEE J. Sel. Top. Quantum Electron. **19(4)**, 1501211 (2013).

P. Crump, C. Frevert, A. Ginolas, S. Knigge, A. Maaßdorf, J. Lotz, W. Fassbender, J. Neukum, J. Körner, T. Töpfer, A. Pranovich, M. Divoky, A. Lucianetti, T. Mocek, K. Ertel, M. De Vido, G. Erbert, and G. Tränkle, *Joule-Class 940-nm Diode Laser Bars for Millisecond Pulse Applications*, IEEE Photonics Technol. Lett. **27(15)**, 1663 (2015).

C. Frevert, P. Crump, F. Bugge, S. Knigge, and G. Erbert, *The impact of low Al-content waveguides on power and efficiency of 9xx nm diode lasers between 200 and 300K*, Semicond. Sci. Technol. **31(2)**, 025003 (2016).

F. Bugge, P. Crump, C. Frevert, S. Knigge, H. Wenzel, G. Erbert, and M. Weyers, *MOVPE growth of laser structures for high-power applications at different ambient temperatures*, J. Cryst. Growth **452**, 258 (2016).

K.H. Hasler, C. Frevert, P. Crump, G. Erbert, and H. Wenzel, *Numerical study of high-power semiconductor lasers for operation at subzero temperatures*, Semicond. Sci. Technol. **32(4)**, 045004 (2017).

M.M. Karow, C. Frevert, R. Platz, S. Knigge, A. Maaßdorf, G. Erbert, and P. Crump, *Efficient 600-W-Laser Bars for Long-Pulse Pump Applications at 940 and 975 nm*, IEEE Photonics Technol. Lett. **29(19)**, 1683 (2017).

Conference contributions

C. Frevert, P. Crump, H. Wenzel, S. Knigge, F. Bugge, and G. Erbert, *Efficiency optimization of high power diode lasers at low temperatures*, European Conf. on Lasers and Electro-Optics (CLEO), 2013, Munich, Germany, May 12-16, poster **CL-P.28-MON** (2013).

P. Crump, C. Frevert, H. Wenzel, F. Bugge, S. Knigge, G. Erbert, and G. Tränkle, *Cryolaser: innovative cryogenic diode laser bars optimized for emerging ultra-high power laser applications*, Conference on Lasers and Electro Optics (CLEO), 2013, San Jose, USA, Jun. 9-14, **JW1J.2** (2013).

C. Frevert, P. Crump, F. Bugge, S. Knigge and G. Erbert, *Study of Waveguide Design for high-power 9xx nm Diode Lasers operating at 200K*, Proc. SPIE **8965**, Photonics West, San Francisco, USA, Feb. 1-6, 89650O (2014).

P. Crump, C. Frevert, H. Hösler, F. Bugge, S. Knigge, W. Pittroff, G. Erbert, and G. Tränkle, *Cryogenic ultra-high power infra-red diode laser bars*, Proc. SPIE **9002**, Photonics West, San Francisco, USA, Feb. 1-6, 900211 (2014).

P. Crump, C. Frevert, F. Bugge, S. Knigge, G. Erbert, G. Tränkle, A. Pietrzak, R. Hülsewede, M. Zorn, J. Sebastian, J. Lotz, W. Fassbender, J. Neukum, J. Körner, J. Hein, and T. Töpfer, *Progress in high-energy-class diode laser pump sources*, Proc. SPIE **9348**, Photonics West, San Francisco, USA, Feb. 07-12, 93480U (2015).

A. Lucianetti, J. Pilar, A. Pranovich, M. Divoky, T. Mocek, K. Ertel, H. Jelinkov, P. Crump, C. Frevert, R. Staske, G. Erbert, and G. Tränkle, *Assessment of high-power kW-class single-diode bars for use in highly efficient pulsed solid-state laser systems*, Proc. SPIE **9348**, Photonics West, San Francisco, USA, Feb. 07-12, 934811 (2015).

C. Frevert, P. Crump, F. Bugge, S. Knigge, A. Ginolas, and G. Erbert, *Low-temperature Optimized 940 nm Diode Laser Bars with 1.98 kW Peak Power at 203 K*, Conference on Lasers and Electro Optics (CLEO), 2015, San Jose, USA, May 10-15, **SM3F.8** (2015).

A. Pietrzak, M. Wölz, R. Hülsewede, M. Zorn, O. Hirsekorn, J. Meusel, A. Kindsvater, M. Schröder, V. Bluemel, J. Sebastian, C. Frevert, F. Bugge, S. Knigge, A. Ginolas, G. Erbert, and P. Crump, *Progress in the development of kilowatt-class diode laser bars for pump applications*, Proc. Advanced Solid State Lasers Conf. (ASSL 2015), Berlin, Germany, Oct. 04-09, **ATh2A.7** (2015).

P. Crump, C. Frevert, A. Ginolas, S. Knigge, A. Maaßdorf, J. Lotz, W. Fassbender, J. Neukum, J. Körner, T. Töpfer, A. Pranovich, M. Divoky, A. Lucianetti, T. Mocek, K. Ertel, M. De Vido, G. Erbert, and G. Tränkle, *Joule-Class 940 nm Diode Laser Bars for Millisecond Pulse Applications*, IEEE Photonics Conference (IPC 2015), Reston, VA, USA, Oct. 4-8, 555 (2015).

C. Frevert, F. Bugge, S. Knigge, A. Ginolas, G. Erbert, and P. Crump, *940nm QCW diode laser bars with 70% efficiency at 1 kW output power at 203K: analysis of remaining limits and path to higher efficiency and power at 200K and 300K*, Proc. SPIE **9733**, Photonics West, San Francisco, USA, Feb. 13-18, 97330L (2016).

Acknowledgements

First and foremost I would like to thank Prof. Dr. Günther Tränkle for giving me the opportunity to write my dissertation at the *Ferdinand-Braun-Institut, Leibniz-Institut für Höchstfrequenztechnik (FBH)* and supervising this work. In frequent discussions, he listened attentively to the progress made, gave valuable feedback and provided new perspectives. He helped me to not get lost in the details, always relating the work to an overarching topic. Any initial doubts about the scientific scope he patiently reasoned away and applied the necessary pressure towards the end of the dissertation. On a personal level, he always offered help to bridge the challenge of raising a family, and performing in the work environment and enriched our meetings with countless memorable anecdotes.

The willingness of Prof. Dr. S. Sweeney and Prof. Dr. M. Kneissl to accept a role as a second and third grader of this work, respectively, is greatly appreciated.

This work would not have been possible without the support and dedication of my colleagues at the FBH. In the long value chain of creating diode lasers, the effort, knowledge and skill of the scientific and technical staff were essential in realizing the optimized designs. The technologists carried out the fabrication of the diode lasers and thus were crucial in transferring ideas into innovative devices which achieved record results. The dissemination of the new findings was greatly aided by my scientific colleagues giving feedback and constructive criticism.

Of the numerous additional people involved, I would specifically like to thank my direct supervisor Dr. Paul Crump for his excellently filled role as a mentor. His vast experience in the field of diode lasers combined with his hands-on, optimistic approach to scientific challenges were of invaluable help, pushing me to seek solutions from new angles. Trusting me with presenting the results of this work at various national and international conferences boosted my confidence and broadened my horizon. He played an essential role in providing highly valued feedback in the revision process. Even in hard times, his British sense of humor livened up the atmosphere and I am grateful for the times spent together at the institute and outside of it.

The experimental part of the work was aided by the broad knowledge of measurement equipment and laboratory experience of Ralf Staske and Dr. Bernd Sumpf who helped design the low temperature measurement station. I am very thankful to Dr. Andre Maaßdorf for providing me with his deep insights into the field of pseudomorphic epitaxial growth and discussing scientific reasons for encountered phenomena. Dr. Götz Erbert's acquired experience and his stock of knowledge helped me avoid scientific dead ends multiple times and his sharp critical thinking polished numerous presentations.

I would like to thank my fellow PhD students at the FBH for enriching my work life with inspiring discussions about their fields of work, sharing their worries and hopes with me and providing distractions during lunch hours. Especially the friendships formed with Jonathan Decker and Martin Winterfeldt created long lasting memories. The reciprocal support during the final stages of completing this work helped tremendously.

The financial support through the project *Cryolaser, SAW-2012-FBH-2* funded within the *Leibniz-Wettbewerb* was immensely appreciated.

Last but certainly not least I want to thank my loving partner Caitlin for her incredible support over the whole course of this work. She always had an open ear for my concerns, gave me the emotional support needed and helped me to focus on finishing the writing process. Our two children, Benjamin and Elisabeth, make me want to excel in life challenges and be an example of a loving, caring and successful father. The time I spend with the kids is repaid to me a thousand times over through their joyfulness, curiosity and exuberance. This work is dedicated to them.

Contents

Chapter 1

Introduction

Motivation

Diode lasers are the most efficient devices in converting electrical into optical energy and are thus key components for numerous modern technologies. They serve as tools for material processing [Li00, Sch00], are used in optical communications and sensing applications, and are vital elements in medical apparatus [Nas14]. Compared to other established laser systems such as gas, solid state or fibre lasers, the direct electrical pumping of diode lasers enables the construction of robust, simple systems with a small footprint and long lifetimes. The variety of available semiconductor materials allows to choose the diode's wavelength from a broad spectrum. The linewidth can be narrowed to < 1 nm by the implementation of gratings [Sch10]. Furthermore, the low cost of diode lasers due to the scalability of the fabrication process makes them commercially appealing. The major disadvantages of high power, high efficiency broad area (BA) diode lasers are their poor beam quality and their low coherence, limiting their use for direct applications in industry. Using them as optical pumps for solid-state and fibre lasers, however, exploits the high powers and efficiencies of the diodes while enabling good beam qualities.

In solid-state laser system with high pulse energies, chirped-pulse amplification (CPA) is widely used to achieve extremely high peak powers. The CPA systems consist of a seed laser whose fs-pulses are temporally stretched (chirped) before being amplified. The resulting ns-pulses can absorb higher pump energies without reaching the damage thresholds of the gain medium, leading to higher amplified pulse energies. After the amplification stages, the pulse is again temporally compressed, so that fs- or ps-pulses with TW-class powers are available [Mas11]. Almost all of the energy in the pulse is thus supplied by the optical power amplifiers, which are made of solid state material and are often cryogenically cooled [Ert11, Fan07]. One common configuration is to use amplifier slabs (e.g. doped YAG crystals) which the laser pulse transverses while being amplified. All the optical energy generated by these amplifiers is provided by optical pumps. The choice of the pumps is often driven by commercial considerations, so flashlamps are commonly used, as they are easily available at a low purchase cost (in $/W).

One of the most prominent and largest example for a CPA based ultra-high energy class laser facility is located in the National Ignition Facility (NIF) of the Lawrence Livermore Laboratories which is performing studies that seek to achieve nuclear fusion by laser ignition [Mil04]. Figure 1.1 schematically shows NIF's layout in which 192 solid-state lasers are focused onto a target within the reaction chamber. Future fusion power plants based on the findings of NIF will require optical pump powers of $\sim 100\,\mathrm{GW}$ [Der11] for Nd:YAG [Mos11] and Yb:YAG [Mas11.2] solid-state amplifiers. In these power plants, the use of flashlamps as pumps is not feasible, as they require high amounts of maintenance and only achieve low efficiency levels at low pulse repetition rates [Ert11]. To ensure a net power generation and to make the laser induced fusion technology affordable, low cost diode lasers with the highest powers and efficiencies are required.

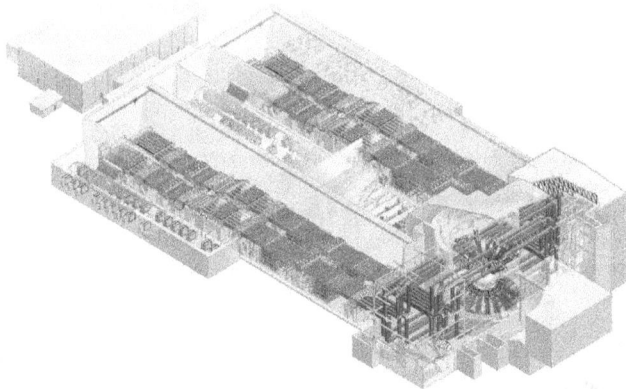

Figure 1.1: Schematic layout of the 192 beamlines and the target chamber at the National Ignition Facility. Figure taken from [NIF16].

Within the *Cryolaser* project[1], the development of these kind of diode laser pump sources was pursued. Specifically, pumps for cryogenically cooled Yb:YAG slabs were targeted, possibly enabling a lower operation temperature of the diode lasers as well. A step size increase in output power and efficiency levels of cryogenically cooled diode lasers is expected to advance their implementation into high energy class laser facilities using solid state laser systems.

State of the art

Power conversion efficiencies of $> 70\%$ have been reported from AlGaAs based, edge-emitting high-power BA lasers operated at room temperature [Kan05, Cru05, Cru13]. Compared to these, commercially available single emitter (SE) BA lasers are usually $w \sim 100\,\mu\mathrm{m}$ wide and emit at wavelengths of $\lambda = 910\text{-}980\,\mathrm{nm}$ with operation output powers $P \geq 10\,\mathrm{W}$ and efficiencies $\eta_E \approx 60\text{-}65\%$ [Lau12]. Output powers of $P = 25\,\mathrm{W}$ have been achieved in continuous wave (CW) mode [Cru09].

---

[1] funded by the Leibniz Gesellschaft under contract SAW-2012-FBH-2

A few studies have been published on the effects of low-temperature operation on the performance of diode lasers. In [Cru06], the authors report an increase of peak efficiency to 85% in a GaAs-based BA laser operated at $-50\,°C$. Significant benefits to output power and efficiency were further reported for InP-based diode lasers at lower operation temperatures in [Mai08], with increases of up to $> +\,300\%$ seen in power and $+\,20\%$ points in efficiency compared to room temperature operation. Record peak conversion efficiencies of 73% have also been published in [Lei10] from InGaAsP-based diode lasers emitting at $\lambda = 1.5\,\mu m$, which were achieved by reducing the operation temperature to 77 K (-196 °C) and using a tailored vertical design.

Electrically operating multiple single emitters on one monolithic piece of semiconductor material in parallel increases the available output power of the component, termed diode laser bar. While such bars have been widely used as quasi-continuous wave (QCW) pump sources for a long time [Die00] and are commercially available with output powers around 300-400 W [Dei08, Ber11, Koh12, Luc15], kW-range bars have also recently been demonstrated, as shown in table 1.1. The chronologically sorted, published results refer to bars with an industrial standard width of 1 cm. It is apparent from the published values in table 1.1, that achievable output powers from a diode laser bar depend heavily on operation conditions (CW / QCW, temperature, ...) and the device's parameters. These parameters will be subsequently discussed for the listed publications and related to the achieved results.

The main difference between CW and QCW operation lies in the amount of dissipated heat in the device, which leads to different cooling requirements. The CW results as well as the early QCW powers of 1.1 kW [Sch07] were achieved from bars mounted on microchannel coolers [Li07, Li08] or similar heatsinks [Kna11] with high heat-removal capacities, while most other publication on QCW bars used simple conduction cooling with lower heat extraction. The need for improved diode lasers to be used as optical pumps is reflected in the higher frequency of publication of QCW bar results starting in 2013 [Kis13, Vet14, Wol14, Che14, Pie15]. Using low duty cycles of $\leq 1\%$, which is the on-time of the laser and is calculated from the pulse width and frequency, the need for expensive and intricate heatsinks with a high heat-removal capacity is avoided and diode laser bars are often integrated into QCW stacked arrays for pumping of solid-state amplifier slabs [Kis13]. The output power of these devices with multiple bars connected in series increases with the number of bars used and represents the need for higher output powers from the optical pumps [Wes13]. While most of the bars were operated at room temperature, a significant increase in output power from 0.7 kW [Li07] to 1 kW [Li08] with an almost constant efficiency is reported which is mainly due to the decreased operation temperature of 8 °C as well as a longer cavity length of $L = 5\,mm$. Wavelength λ and fill factor (FF), which describes the fraction of the 1 cm-wide bar occupied by emitters, are roughly the same. A good overview about impact of the lateral design choice (FF and L) of high power diode laser bars on power and efficiency as well as the underlying economic considerations has been presented in [Cru14]. The exceptionally high powers of 1.8 kW quoted in [Che14] were obtained by using a multi-junction vertical design in which laser structures are epitaxially grown on top of each other, separated by tunnel junctions. While all other cited publications employ single junctions, the use of the three junctions in [Che14] in one bar lead to a strong increase in output power. However, efficiencies are not given in the publication and are assumed to be lower than for the equivalent use of three single junction bars.

Table 1.1: Published kW-class diode laser bar results with given operation conditions and device parameters.

publication	year	operation	temperature [°C]	λ [nm]	FF [%]	L [mm]	power [kW]	efficiency [%]
[Sch07]	2007	QCW	15	980	44	4	1.1	< 35
[Li07]	2007	CW	25	940	77	3	0.7	58
[Li08]	2008	CW	8	940	83	5	1.0	56
[Kna11]	2011	CW	20	940	77	5	0.94	55
[Kis13]	2013	QCW	20	940	75	1.5	0.71	
[Vet14]	2014	QCW	25	940			0.55	50
[Wol14]	2014	QCW	25	940	75	1.5	0.6	52
[Che14]	2014	QCW	10	870	80	3	1.8	
[Pie15]	2015	QCW	25	940	50	4	0.8	60

No further studies have been published regarding the use of cryogenic operating temperatures for high power single emitters and bars. The benefit of reduced operating temperatures to the output power of bars presented in [Li08] combined with the efficiency increases observed in single emitters at cryogenic temperatures [Cru06] encourages the development of high power, high efficiency diode laser bars for temperatures of 200 K.

Goal of this work

The technical goal of this work is to design high power, high efficiency QCW diode laser bars for peak performance at an operation temperature of $\sim$ 200 K. This relies on the understanding of the changing semiconductor properties at low temperatures and how they affect the electro-optical properties of the lasers at reduced operation temperature. The temperature dependence of the factors that limit optical output power and power conversion efficiency at high powers needs to be studied and their relative importance has to be assessed. The impact of the vertical and lateral laser design on the performance at lower temperatures has to be analyzed starting on single emitter level. Revised and low temperature optimized vertical designs are sought, which can be fabricated into QCW diode laser bars with the highest output powers and efficiencies.

Specifically, the development and fabrication of exemplary QCW diode laser bars with output powers > 1.5 kW and efficiencies > 60% at this point are targeted, operated at a temperature of $\sim$ 200 K. Their specifications regarding amongst others wavelength and operation conditions, are governed by their intended use as pump sources for cryogenically cooled Yb:YAG solid state amplifiers and are listed in table 1.2.

Table 1.2: Targeted specifications for low temperature optimized high power, high efficiency QCW diode laser bars.

parameter	symbol	
heatsink temperature	T_{HS}	$\sim 200\,\text{K}$
frequency	f	$10\,\text{Hz}$
pulse length	τ	$1.2\,\text{ms}$
wavelength	λ	$940\,\text{nm}$
bar width		$1\,\text{cm}$
output power	P	$> 1.5\,\text{kW}$
efficiency at P	$\eta_E >$	60%

For economic considerations, the cavity length of the bar should be as short as possible without compromising the power and efficiency levels. Similarly, conduction cooled packaging of the bars instead of microchannel coolers is desired to limit the system complexity. While low spectral width and low vertical beam divergence are potentially beneficial for the pumping of the solid state amplifiers [Ert11], they are not targeted in this work. Wavelength stabilization can be implemented into developed structures at a later stage, for example by surface etched gratings [Fri12].

In order to achieve the high powers levels, the use of high FF ($> 70\%$) in the bars is essential. Assuming a FF = 75% bar (schematically drawn in figure 1.2) comprised of BA lasers with a usual $w = 100\,\mu\text{m}$ and $L = 4\,\text{mm}$, this represents an output power of $P > 20\,\text{W}$ per single emitter. This estimation does not take into account any deterioration of single emitter performance on bar level due to mounting stress or cross talk between the emitters. However, as long term operation at powers up to $20\,\text{W}$ of single emitters has been demonstrated in CW trials [Cru09], QCW bar operation at the same power level should be feasible.

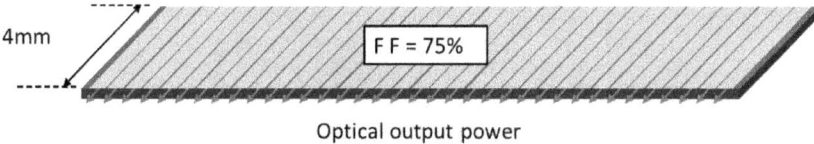

Figure 1.2: Schematic lateral layout of a high fill factor bar.

The work is structured as follows: first, the basic principles of high power, high efficiency diode lasers are explained in chapter 2. The parameters determining optical output power and power conversion efficiency are deduced and their importance for high power, high efficiency operation is estimated. Furthermore, a brief summary of the changes at low temperatures of relevant semiconductor properties is given. In chapter 3, the experimental setup used to characterize the electro-optical performance of single emitters and bars at room temperature as well as lower temperatures is described. The design of a low temperature measurement station and the established accurate power calibration are laid out. Chapter 4 begins with a benchmark assessment of single emitters of different established vertical designs, in which the impact of various vertical design parameters on the electro-optical performance at lower temperatures is studied. High power single emitters and bars of two of the vertical designs are measured and their efficiency-limiting factors are analyzed. In chapter 5, the option of reducing the main efficiency limiting factor, the series resistance, by lowering the Al-content in the waveguide is explored. Low-current measurements are performed on single emitters of epitaxial designs using various Al-contents and insights on the temperature dependence of electrical and optical parameters are gained. Record output powers and efficiencies of fabricated low temperature bars using low Al-content waveguides are presented and options to further increase efficiency at the high powers are discussed. Chapter 6 contains a further approach to reduce series resistance by decreasing the p-side waveguide thickness of the vertical structure. Combined with low Al-content waveguides, low temperature optimized epitaxial designs are deduced and experimentally assessed in bar and single emitter format. Finally, all results are summarized and a brief outlook to further improvements is given.

Chapter 2

Theoretical background of high power, high efficiency diode lasers

In this chapter, the basic principles of high power diode lasers are laid out. Diode lasers have been extensively studied and numerous books have been written on the topic. Here, a brief summary of the underlying physics and the properties of the employed materials is given, before diode laser properties and parameters are selectively introduced. The discourse follows the introduction of diode lasers in [Col98] and uses figures from within. No claim for completeness is made as it is out of the scope of this work and the author encourages further reading. Specifically, a comprehensive introduction into solid state physics is given in [Sze01], while the material system of AlGaAs diode lasers is extensively studied in [Ada93]. For additional information on the working principles of diode lasers, [Col98] is recommended, while the device engineering and the fabrication necessary for high power devices is well described in [Die00].

The theoretical background provided in this chapter is limited to topics which are crucial to understand the effects limiting the output power and conversion efficiency in AlGaAs diode lasers. In the first part, the basic working principles of AlGaAs / InGaAs based high power diode lasers are introduced, followed by a discussion of important laser parameters including threshold, output power and power conversion efficiency. Based on a numerical model, the electro-optical performance of a typical single emitter diode laser is simulated and compared for varying characteristic parameters. The second part covers the effect of low temperature operation on the diode laser properties and puts an emphasis on the expected increase of bulk conductivity.

2.1 Basic principles of AlGaAs / InGaAs based diode lasers

Unlike most other laser systems, in which the gain medium has to be optically pumped, diode lasers can be directly pumped by an electrical current, which allows much more efficient operation. Stimulated recombination of carriers between the conduction and valence band of the semiconductor gain medium leads to the emission of photons with energies slightly larger than the band gap. A simple vertical setup of a diode laser is the double-heterostructure which uses a thin layer of undoped active material sandwiched between n- and p-doped cladding layers with a larger band gap. By applying a forward bias, the potential well of the undoped region is filled with injected holes and electrons from the p- and n-side, respectively, and the captured carriers recombine in the active region / zone. The necessary optical feedback for completing a laser is established by cleaved facets of the semiconductor material, which provide a large refractive index step at the semiconductor-air boundary and photons are reflected back into the semiconductor. A transverse dielectric optical waveguide (x direction) is formed by the higher index of refraction in the n- and p-doped cladding layers, so that the light propagates along the plane of the undoped active region (z direction) between the facets for edge emitting diode lasers as schematically drawn in figure 2.1 (a). Separating the optical confinement of photons from the electrical confinement of the carriers in structures termed separate confinement heterostructures (SCH) and using a thin ($d \sim$ 5-10 nm) quantum well (QW) leads to more efficient operation of the diode laser. In figure 2.1 (b), the band gap profiles as well as the electrical field both in transverse direction of such a SCH are shown.

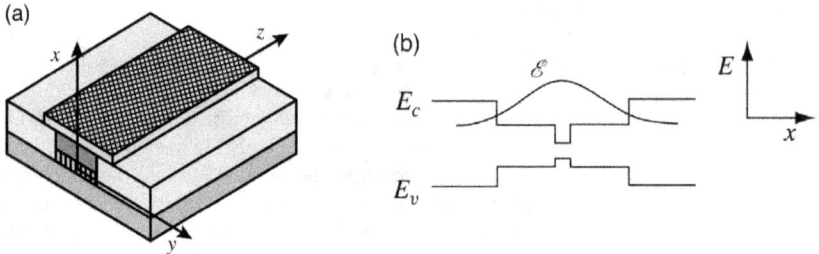

Figure 2.1: (a) Schematic setup of an edge emitting diode laser. (b) Transverse band structures for a standard separate confinement heterostructure. Figures taken from [Col98].

In the structure displayed in figure 2.1 (b), the outer layers (also termed cladding layers) confine the photons to the inner layers (termed waveguides), which in turn provide a confinement for the electrical carriers in the QW. The requirements for the semiconductor materials with different band gaps used to fabricate SCH diode lasers include a common crystal structure and a similar lattice constant for defect-free epitaxial growth of single-crystal layers on top of each other. An exception to the demand for uniform lattice constants is the quantum well. Here, a small lattice mismatch between the QW layer and the surrounding waveguide layers is often desired. The strain in the quantum well causes a separation of the valence band into two bands for light and heavy holes. The former provides optical gain for electromagnetic waves in transversal *magnetic* polarization with the vertical field vector perpendicular to the plane of the QW, while the latter leads to gain for transversal *electric* polarization, in which the vertical field vector is parallel to the QW. Furthermore, lasing threshold and optical loss are reduced in strained layer QWs. Diode lasers in the 0.7-1.6 µm wavelength range most commonly consist of III-V compounds with a direct gap, in which momentum conservation without phonon contribution in the carrier recombination process is satisfied. In figure 2.2, the bandgap of several III-V semiconductors is plotted against the lattice constant. Immediately, the AlGaAs ternary line stands out, as it is almost vertical. This means, that the lattice constant remains approximately unchanged when substituting Al for Ga in GaAs. Thus, a wide range of bandgaps is accessible for designing SCH diode lasers by using layers with different $Al_xGa_{1-x}As$ compositions. For diode lasers in the 9xx nm wavelength range, InGaAs is used as a material for the strained QWs.

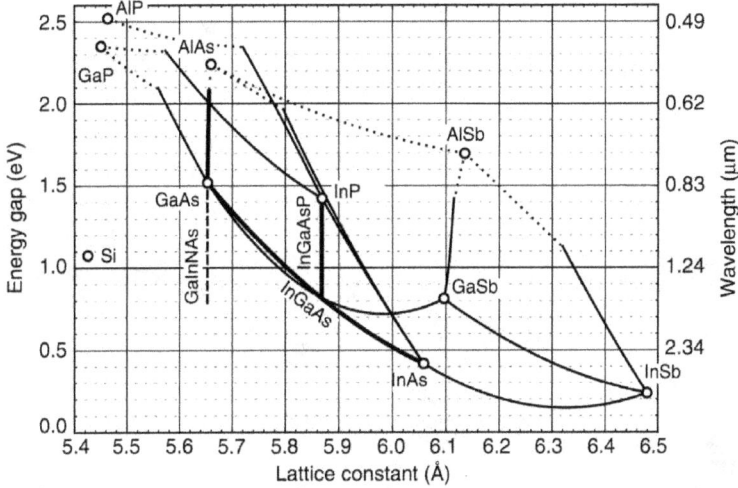

Figure 2.2: Bandgap energy plotted against lattice constant for III-V semiconductors. The single points are binary compounds, with the curves representing ternary compounds with direct (solid) and indirect (dashed) bandgaps. Figure taken from [Col98].

The deposition of the multiple single-crystal lattice-matched layers is achieved by epitaxial growth over a suitable substrate. For edge emitting lasers, the most common deposition techniques are the metal-organic chemical vapor deposition (MOCVD) and molecular beam epitaxy (MBE). While MOCVD growth can remove surface damages of the substrate and offers high quality interfaces between the different layers, MBE excels in film uniformity and well defined thicknesses. For the diode lasers discussed in this work, $Al_xGa_{1-x}As$ layers for the cladding and waveguide layers are deposited by MOCVD on GaAs substrates, with the gain provided by one or multiple strained InGaAs QWs.

After the epitaxial growth, the wafer with the vertical laser structure is laterally patterned to provide lateral optical and current confinement. In this wafer process, the current's path is restricted for edge-emitting diode lasers by applying narrow metallized contact stripes to the top-side (p-side) of the grown structure. Together with insulating layers, selective etching and / or implantation insolation of p-regions, this limits the lateral carrier distribution on the vertical path through the device. In turn, the creation of photons is also laterally confined to the areas accessible by the current. Possible further implementation of lateral index steps provides boundaries for the extent of the optical field guiding the photons along the axis of the cavity. For common high power broad area lasers (BAL) which are discussed in this work, current is limited to a stripe of $w = 70\text{-}200\,\mu\text{m}$ width. A typical BAL single emitter with its dimension is displayed in figure 2.3.

After the lateral wafer process, diode bars containing single emitters are cleaved from the wafer with determined cavity lengths. The facet are cleaned, passivated and coated to reflect the light back into the cavity (rear facet with high reflectivity R_r) and transmit the optical output (front facet with anti reflection coating R_f). Finally, single emitters are cleaved from the bar and mounted to expansion matched CuW heatsinks, establishing electrical connections and providing efficient heat removal.

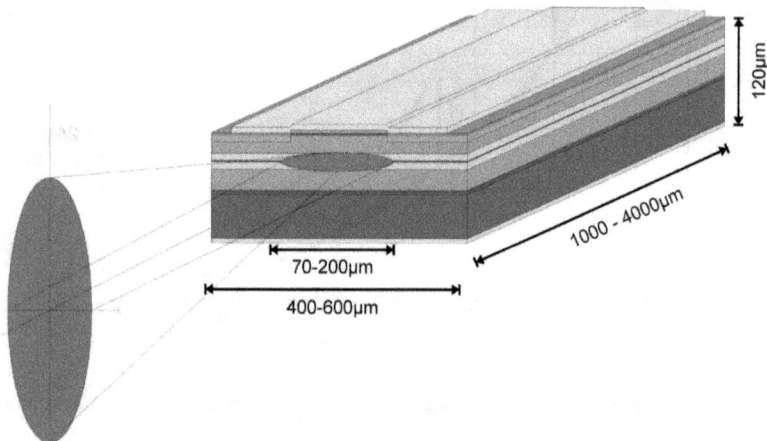

Figure 2.3: Schematics and dimensions of a high power single emitter broad area laser.

2.1.1 Lasing threshold

The conditions at the onset of lasing, namely the threshold gain and the threshold current, are subsequently derived for edge emitting Fabry-Perot (FP) diode lasers. The propagating optical wave along the laser's cavity is described by a normalized transverse field profile $U(x, y)$ established by the dielectric waveguide and its propagation can be expressed by $\exp(-i\beta z)$ with β being the complex propagation constant. For the time- and space-dependent electrical field, polarized in TE (y) direction, this leads to

$$\vec{E}(x, y, z) = E_0 \cdot U(x, y) \exp\left(i\left(\omega t - \beta z\right)\right)\vec{e}_y \qquad (2.1)$$

with E_0 being the field's magnitude. β can be expressed as a function of wavelength λ, the transverse modal gain Γg_{mod} and loss α_i and the effective index of refraction for the mode n_{eff}

$$\beta = \frac{2\pi n_{\mathrm{eff}}}{\lambda} + \frac{i}{2}\left(\Gamma g_{\mathrm{mod}} + \alpha_i\right). \tag{2.2}$$

The transverse confinement factor Γ describes the overlap of the optical field with the active zone. At lasing threshold, the electrical field of the lasing mode after one round-trip through the cavity of length L is exactly replicated, $\vec{E}\left(z = 0\right) = \vec{E}\left(z = 2L\right)$. At this point, the mode's field has been reflected at the facets with $r_{\mathrm{f}} = \sqrt{R_{\mathrm{f}}}$ and $r_{\mathrm{r}} = \sqrt{R_{\mathrm{r}}}$.

$$\vec{E}\left(z = 0\right) = r_{\mathrm{f}} r_{\mathrm{r}} \vec{E}\left(z = 2L\right)\big|_{\Gamma g_{\mathrm{th}}} \tag{2.3}$$

The real part of equation 2.3, e.g. its magnitude, can be separately analyzed and leads to the threshold gain needed to compensate for the mirror and internal losses, α_{m} and α_{i}, respectively

$$\Gamma g_{\mathrm{th}} = \alpha_{\mathrm{m}} + \alpha_{\mathrm{i}} \tag{2.4}$$

with the mirror loss

$$\alpha_{\mathrm{m}} = \frac{1}{L}\ln\left(\frac{1}{r_{\mathrm{f}} r_{\mathrm{r}}}\right) = -\frac{1}{2L}\ln\left(R_{\mathrm{f}} R_{\mathrm{r}}\right). \tag{2.5}$$

The current needed to establish threshold I_{th}, e.g. lead to a modal gain satisfying equation 2.4, is derived from the gain g, which can be approximated by a logarithmic function of the carrier density N as described in [DeT93]

$$g = g_0 \ln\left(\frac{N}{N_{\mathrm{T}}}\right) \tag{2.6}$$

in which g_0 the material gain and N_{T} the transparency carrier density. As the carrier density N is established by the current density J, it can be approximated linearly as $N \approx J$, leading to

$$g \approx g_0 \ln\left(\frac{J}{J_{\mathrm{T}}}\right) \tag{2.7}$$

with J_{T} the transparency current density. At threshold, this can be re-written as a function of threshold current density J_{th} using equation 2.4

$$J_{\mathrm{th}} = J_{\mathrm{T}} \exp\left(\frac{\alpha_{\mathrm{i}} + \alpha_{\mathrm{m}}}{\Gamma g_0}\right) \tag{2.8}$$

To calculate the threshold current in BALs, the threshold current density has to be multiplied by the pumped area, given by the contact length (mostly equivalent to the laser cavity length L) and the contact width w. The variables J_{T}, α_{i} and Γg_0 are three out of four internal parameters dependent on the epitaxial design and their values are essential for assessing the quality of a vertical structure. The magnitudes of J_{T} and Γg_0 are commonly found by determining the threshold current of FP lasers of the same vertical design with different mirror loss. The latter is achieved by cleaving diodes with different cavity lengths without coating the facets (for a universal $R = R_f = R_r \approx 30\%$). The logarithmic threshold current density can then be plotted as a linear function of inverse cavity length

$$\ln\left(J_{\text{th}}\right) = \ln\left(\frac{I_{\text{th}}}{w \cdot L}\right) = B + A \cdot \frac{1}{L} = \left[\ln J_{\text{T}} + \frac{\alpha_i}{\Gamma g_0}\right] + \left[\frac{\ln 1/R}{\Gamma g_0}\right] \cdot \frac{1}{L}. \tag{2.9}$$

A linear fit to the measurement values gives numbers for the intersect with the ordinate B and the slope of the fit A, from which the values for J_{T} and Γg_0 can be calculated.

2.1.2 Optical output power

Above threshold, most of the optical output power P of a diode laser is generated in the active gain medium, e.g. the QW, by the stimulated recombination of electrons and holes. Within the undoped quantum well, charge neutrality prevails, e.g. the electron density is equal to the hole density. Setting up a rate equation for the carrier density N in the QW, the injection rate G_{inj} of carriers into the QW and the rate of recombination R_{rec} of carriers can be linked

$$\frac{dN}{dt} = G_{\text{inj}} - R_{\text{rec}} = \frac{\eta_i I}{qV} - \frac{N}{\tau} - R_{\text{st}} \tag{2.10}$$

with η_i the internal differential efficiency, q the unit charge, I the applied current, V the volume of the active region, τ the carrier lifetime due to spontaneous and non-radiative recombination and carrier leakages out of the QW and R_{st} the stimulated recombination rate. A similar rate equation of generation and loss terms can be formulated for the photon density N_{ph}, as described in detail in [Col98]. For lasers operated in continuous wave (CW) mode, a steady state is reached above threshold where $dN/dt = dN_{\text{ph}}/dt = 0$. This allows the combination of the two rate equations for N and N_{ph} and gives the steady-state photon density

$$N_{\text{ph}} = \frac{\eta_i \left(I - I_{\text{th}}\right)}{qv_g g_{\text{th}} V} \tag{2.11}$$

where v_g is the group velocity of the cavity mode. The optical output power is proportional to N_{ph}. The output power P from the mirrors is gained by multiplying the optical energy stored in the cavity, given by $E_{\text{os}} = N_{\text{ph}} h\nu V_{\text{ph}}$, where $h\nu$ the energy per photon and V_{ph} the cavity volume occupied by the photons, by the energy loss rate through the mirrors $1/\tau_m = v_g \alpha_m$:

$$P = v_g \alpha_m N_{\text{ph}} h\nu V_{\text{ph}}. \tag{2.12}$$

By using equations 2.11 and 2.4, the output power above threshold simplifies to

$$P = \eta_d \frac{hc}{q\lambda}(I - I_{\text{th}}) \tag{2.13}$$

where the wavelength independent external differential efficiency η_d is given by

$$\eta_d = \frac{\eta_i \alpha_m}{\alpha_i + \alpha_m} = \frac{q\lambda}{hc} \cdot dP/dI. \tag{2.14}$$

Contained in η_d is the missing internal parameter η_i as well as the internal parameter α_i. Similar to the determination of the internal parameters J_T and Γg_0, η_i and α_i can be accessed from length dependent measurements. By measuring the inverse of the slope S, given by $S = dP/dI$ and correcting for the wavelength, the inverse of the external differential efficiency is plotted for different cavity lengths. Rearranging equation 2.14, the mirror loss dependence of η_d can be transferred into a dependence on cavity length

$$\frac{1}{\eta_d} = D + C \cdot L = \left[\frac{1}{\eta_i}\right] + \left[\frac{\alpha_i}{\eta_i \ln(1/R)}\right] \cdot L. \tag{2.15}$$

Linear fitting of the measurement points gives fit values for the intercept with the ordinate (D) and the slope of the fit (C) from which η_i and α_i can be calculated.

2.1.3 Electro-optical conversion efficiency

One of the most important parameters in high power diode lasers is their electro-optical power conversion efficiency η_E, also termed wall plug efficiency. It states the percentage of the electrical input power P_{el} supplied to the device that is transformed into optical output power P:

$$\eta_E = \frac{P}{P_{el}} = \frac{P}{U \cdot I} \tag{2.16}$$

where U is the applied bias voltage at a given current. For all diode laser designs, device engineers strive to increase η_E at a point of operation to enhance optical output power, mitigate circuit heating and reduce system power requirements. In order to achieve higher efficiencies, the optical power has to be increased or the necessary electrical power has to be reduced. Focusing on the latter, the current dependent voltage can be linearly approximated by

$$U(I) = U_0 + R_S I \tag{2.17}$$

with U_0 the turn-on voltage and R_S the electrical series resistance. This simple representation of U assumes a uniform, current-independent series resistance is valid for simple *pin*-junctions [Sze01]. The validity of this assumptions in the diode lasers driven to high currents will have to be assessed and one challenge within this work is to identify vertical design choices leading to different current-dependences of series resistance. U_0 represents the minimal voltage required to achieve population inversion. It consists of two contributions $U_0 = U_{gap} + U_{d,0}$: the voltage difference between the quasi Fermi-levels in the quantum well U_{gap} needed for the bandgap transition (at $\lambda = 975$ nm, $U_{gap} = 1.272$ V) and an additional voltage drop $U_{d,0}$. As this voltage drop can be calculated from the intercept of a linear fit to the voltage-current curve and the knowledge of the band gap, e.g. photon energy / wavelength, it is termed defect voltage (at 0 A). It originates from non-ideal interfaces and the additional bias due to the differences in band gap between cladding and QW material with the highest and lower band gap energies, respectively. Equation 2.16 can be further analyzed by substituting P from equation 2.13 and U from equation 2.17 and separating it into four contributions:

$$\eta_E = \eta_i \times \left[\frac{\alpha_m}{\alpha_m + \alpha_i}\right] \times \left[\frac{h\nu}{q(U_0 + R_S I)}\right] \times \left[\frac{I - I_{th}}{I}\right]. \tag{2.18}$$

The first factor, η_i, describes the efficiency at which terminal current generates carriers in the active region and is typically close to 1 for high power, high efficiency devices. Carriers of the input current lost in the vertical layer structure on the way to the quantum well due to leakage currents will decrease η_i. The second factor is a measure for the optical cavity's efficiency, reduced by optical absorption of photons within the laser (α_i). The factor is reduced for low mirror loss and high internal loss and is around 0.9 for the discussed lasers. The third factor represents the electrical loss due to voltage drops within the device. Its magnitude depends less on the fixed contribution of U_0 but strongly on the current. Especially for high currents, this factor will become the main limiting factor for higher η_E (depending on the magnitude of R_S). Finally, the last factor is a measure for the excess current which does not contribute to output power (namely the current necessary to establish threshold). With increasing current, this factor becomes less and less important and for the current levels (and threshold currents) anticipated in this work, it will tend to 1.

2.1.4 Waste heat

Since in practice the power conversion efficiency of diode lasers is always < 1, the electrical power not transformed into optical power (due to the loss factors described above) stays in the device as waste heat

$$P_{\text{waste}} = U \cdot I - P \tag{2.19}$$

and causes a temperature increase, which has to be countered by thermal management of the diode laser. On a microscopic level, a rise in the semiconductor temperature T changes the Fermi-Dirac distribution for energy E

$$f(E) = \frac{1}{\exp\left[(E - E_{\text{F}})/k_{\text{B}}T\right] + 1} \tag{2.20}$$

with E_{F} the Fermi level and k_{B} the Boltzmann constant. This leads to the occupation of higher energy states in the quantum well, effectively lowering the confinement of the carriers, increasing transparency carrier density and broadening the gain spectrum. The reduced confinement decreases η_i and the leaked carriers combined with the broader gain spectrum and the higher carrier density at threshold lead to higher α_i. Furthermore, the energy difference between the Fermi-levels in conduction and valence band decreases with increasing temperature. Subsequently, photon energy is reduced and the wavelength experiences a red-shift. Based on these considerations on a microscopic scale, the macroscopic laser parameters threshold current and the external differential efficiency will be altered. An increasing temperature will increase the threshold because of the broadened gain, increased N_{T} and the potentially higher internal loss, while the lower η_i and higher α_i will decrease η_{d}.

The threshold current shows an exponential increase with temperature and a common approximation of I_{th} for a certain increase in temperature ΔT is given by

$$I_{\text{th}}(T + \Delta T) = I_{\text{th}}(T) \cdot \exp(\Delta T/T_0) \tag{2.21}$$

with T_0 a temperature constant. Similarly, the temperature dependence of the external differential efficiency can be written as an exponential decay

$$\eta_d\,(T + \Delta T) = \eta_d(T) \cdot \exp\left(-\Delta T/T_1\right) \qquad (2.22)$$

where T_1 is a temperature constant. The values for T_0 and T_1 are important characteristics for a vertical layer design. Laser structures with higher T_0 and T_1 values are less affected by temperature changes (for example caused by the dissipated heat or a change in heatsink temperature). Typical diode lasers in the 9xx nm wavelength range, optimized for high power, high efficiency operation show values of $T_0 = 80\ldots150\,\mathrm{K}$ and $T_1 > 500\,\mathrm{K}$ [Kau18]. While it is difficult to determine T_0 and T_1 for a given vertical structure theoretically, these parameters can be obtained by measuring the pulsed light-current, e.g. power-current (PI) curves of processed diode lasers at different heatsink temperatures. Plotting threshold current and external differential efficiency versus heatsink temperature, exponential fits following equations 2.21 and 2.22 are applied with T_0 and T_1 as fit values.

The waste heat of equation 2.19 is generated within the laser device, mostly in or close to the QW. As the heatsink with temperature T_{HS} is at a different vertical position than the heat generation, heat flux through the device has to overcome a thermal resistance R_{th}. This leads to a temperature difference ΔT_{AZ} between active region and heatsink, given by

$$\Delta T_{\mathrm{AZ}} = R_{\mathrm{th}} \cdot P_{\mathrm{waste}}. \qquad (2.23)$$

The lower R_{th}, the better the cooling of laser diode becomes. As the vertical distance of the surface to the active zone of a vertical structure is generally smaller for the p-side than for the n-side (due to the substrate, which is on the n-side), diode lasers are mounted p-side down (if possible) to achieve smaller values of R_{th}.

2.1.5 Influence of laser parameters on output power and efficiency

The laser parameters introduced in the previous sections influence the performance of the assembled diodes. Based on the exponential behavior of threshold current and external differential efficiency, the output power for a certain CW drive current can be calculated using [Die00]

$$\begin{aligned} P &= \frac{hc}{q\lambda} \cdot \eta_d \exp\left\{[-R_{\mathrm{th}}\left((U_0 + R_S I) - P\right]/T_1\right\} \\ &\quad \times \left[I - I_{\mathrm{th}} \exp\left\{[R_{\mathrm{th}}\left((U_0 + R_S I) - P)\right]/T_0\right\}\right]. \end{aligned} \qquad (2.24)$$

Together with the assumption of a linear voltage-current characteristic (equation 2.17), the electro-optical characteristics of a laser with a given set of parameters can be extra-polated. Changing the values of specific parameters, their influence on the optical output power and power conversion efficiency can be visualized.

For broad area diode lasers later discussed in chapters 5 and 6, common values for the laser parameters are given in table 2.1. Here, the typical values are listed as well as maximal and minimal values.

Table 2.1: Laser parameters of an exemplary laser diode. Given are typical values as well as a range.

parameter	notation	unit	min. val.	typ. val.	max. val.
length	L	[µm]	3000	4000	6000
width	w	[µm]	90	100	200
wavelength	λ	[nm]	940	970	980
transp. curr. density	J_T	[A/cm^2]	70	90	130
modal gain	Γg_0	[cm^{-1}]	6	8	10
internal loss	α_i	[cm^{-1}]	0.3	0.4	0.6
mirror loss	α_m	[cm^{-1}]	4	5	6
int. diff. efficiency	η_i	[%]	85	90	95
series resistance	R_S	[mΩ]	10	20	30
defect volt. at 0 A	$U_{d,0}$	[mV]	10	20	30
thermal resistance	R_{th}	[K/W]	2	3	4
temp. constant for I_{th}	T_0	[K]	70	100	150
temp. constant for η_d	T_1	[K]	300	500	800

In table 2.1, the bottom set of parameters is analyzed as follows. First, the effect of one specific parameter on power and efficiency is explored by using its typical as well as its minimal and maximal value in equation 2.24, while keeping the other parameters at their typical value. The three resulting PUI and efficiency characteristics for the typical, minimal and maximal value are plotted in figures 2.4, 2.5 and 2.6 up to a current of $I = 20\,$A for each laser parameter listed. Subsequently, the variables in equation 2.24 are traced back to the analyzed laser parameters and are assessed for their impact on high powers and high efficiencies. It should be noted, though, that high power phenomena such as longitudinal spectral hole burning, two photon absorption or band bending effects are *not* included in the extrapolation.

Looking at figures 2.4-2.6, the most important parameter for reaching high output powers with high efficiencies is the slope. Mainly, a high internal differential efficiency (figure 2.4 (a)) is crucial, as an increased η_i directly translates to an increased η_E. Low internal loss (figure 2.4 (b)) and high mirror loss (figure 2.5 (b)) also lead to an increased slope.

Not primarily affecting the output power but greatly reducing power conversion efficiency at high currents is the bias voltage. While $U_{d,0}$ has only a negligible effect (figure 2.6 (b)), a high series resistance strongly decreases η_E (figure 2.6 (a)). Due to the higher amounts of dissipated heat, the power is in turn degraded by thermal roll-over.

Thermal properties such as a high thermal resistance or a high temperature sensitivity of the slope (figures 2.1 (a) and 2.6 (d)) reduce output power at high currents due to the larger amounts of waste heat remaining in the structure. The temperature sensitivity of the threshold (figure 2.6 (c)) does not strongly affect power or efficiency for the used values.

Finally, the threshold current has a significant impact on the maximum efficiency, which is reached at lower currents. At higher currents, however, a higher threshold due to an increased transparency current density or a lower modal gain (figures 2.4 (c) and (d)) plays only a marginal role on the output power and efficiency.

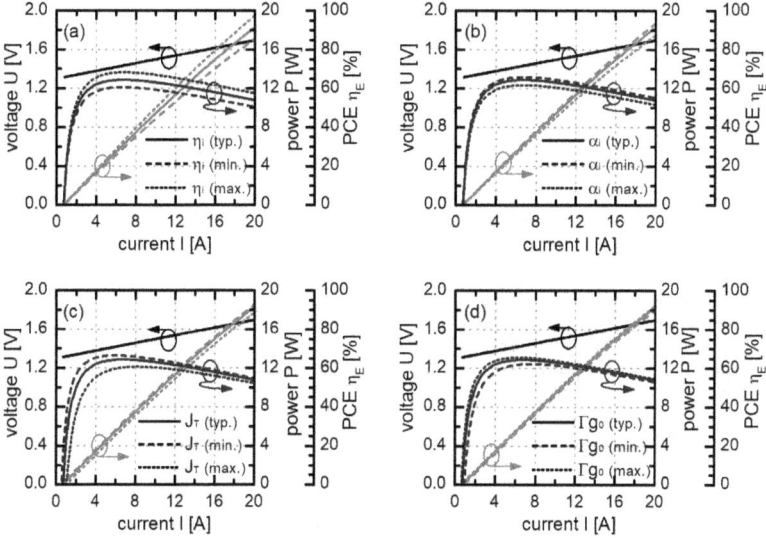

Figure 2.4: Power, voltage and efficiency extrapolation for the different values of the internal parameters (a) internal differential efficiency, (b) internal loss, (c) transparency current density and (d) modal gain. The minimal, typical and maximal values for an exemplary BAL are given in table 2.1.

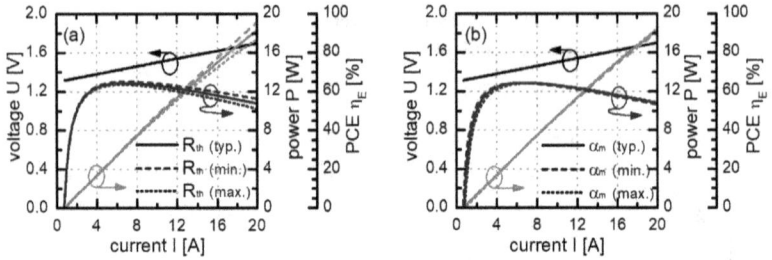

Figure 2.5: Power, voltage and efficiency extrapolation for the different values of the external parameters (a) thermal resistance and (b) mirror loss. The minimal, typical and maximal values for an exemplary BAL are given in table 2.1.

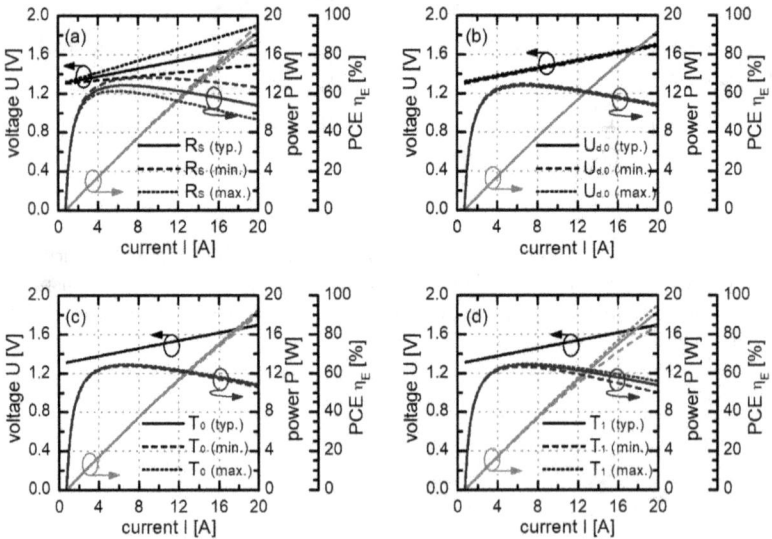

Figure 2.6: Power, voltage and efficiency extrapolation for the different values of the characteristic parameters (a) series resistance, (b) defect voltage at 0 A, (c) temperature constant for threshold and (d) temperature constant for external differential efficiency. The minimal, typical and maximal values for an exemplary BAL are given in table 2.1.

2.2 Semiconductor properties at reduced temperatures

While the majority of diode laser research focuses on the optimization at room temperatures or elevated temperatures, this work addresses an operation temperature of $\sim 200\,\mathrm{K}$. Therefore it is necessary to revise the effect of lower temperatures on the semiconductor properties. First, the changes in the active zone are discussed, affecting the optical behavior. Subsequently, the impact of reduced phonon number and lower thermal energy of the carriers on the semiconductor's conductivity is assessed. Measured bulk conductivity in grown samples proves the reduction of series resistance at lower temperatures.

The density of filled states in the QW is given by the density of states for a 2D system $\rho(E)$ and the Fermi-Dirac-distribution (FD). While the density function $\rho(E)$ does not have a direct temperature dependence, the FD distribution changes with temperature according to equation 2.20. Incremental changes to the semiconductor temperature have been discussed previously. Here, the effect of drastically lowered operation temperature is explored. For room temperature and the temperature of interest for this work, 200 K, the FD distribution is plotted in figure 2.7 for the relative energy E/E_F. The flanks at room temperature extend to higher energies, while in the low temperature case the distribution reaches its maximum and minimum value for smaller and higher E/E_F, respectively. This means, that for a fixed density function, carriers will be spread over a lower energy range, leading to a narrower carrier distribution in E and a higher gain for QW diode lasers. Simultaneously, the band gap increases with decreasing temperature, phenomenologically described by the Varshni relation. This leads to a blue-shift in wavelength.

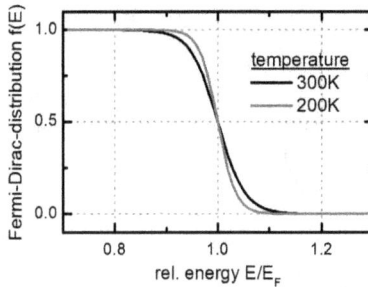

Figure 2.7: Fermi-Dirac distribution plotted against relative energy for temperatures of 200 K and 300 K.

The narrowing of the occupied energy range also benefits the carrier confinement within the QW. The population probability is increased for low energy levels, so that thermal leakage of carriers out of the QW is reduced. The effect of a broader FD distribution on the leakage current out of the active region is schematically depicted for electrons escaping into the p-side in figure 2.8. However, not all carriers with higher energy levels than the barrier are automatically lost due to leakage, as the interface causes a reflection of carriers. The actual number of carriers thermally escaping the QW are further influenced

19

by factors such as their randomly distributed velocities and drift and diffusion processes on the edge of the active region. Generally, any leakage current from the QW will decrease the internal differential efficiency, as carriers are no longer available for the intended stimulated recombination.

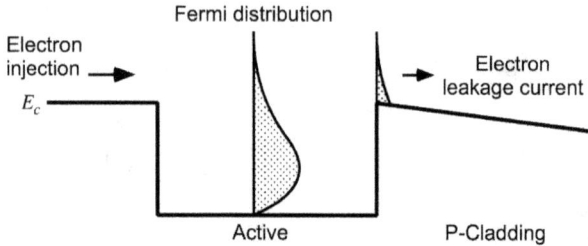

Figure 2.8: Schematic depiction of carrier leakage from the active region due to a broad Fermi-Dirac distribution. Figure taken from [Col98].

2.2.1 Bulk conductivity at lower temperatures

Recognizing that resistance ρ is reciprocal to the conductivity σ of a bulk semiconductor layer

$$\sigma = 1/\rho = e\left(n\mu_{\mathrm{e}} + p\mu_{\mathrm{h}}\right) \tag{2.25}$$

the temperature dependence of the resistance can be gained by analyzing the free carrier concentration n, p and the electron and hole mobilities $\mu_{\mathrm{e,h}}$. Phenomenologically, the temperature dependence of these four variables follows the behavior shown in figure 2.9 as explained in [Sze01].

In figure 2.9 (a), three regions of carrier concentration in a doped semiconductor can be identified. Starting at low temperatures (high T^{-1}) in the freeze-out range, more and more carriers with $E_{\mathrm{th}} = k_{\mathrm{B}}T$ are thermally activated when the temperature is increased. Above a certain temperature, all doping centers are ionized and no further increase in carrier concentration is achieved by increasing the temperature (saturation range). Finally, though, the temperature is high enough to excite carriers across the band gap and intrinsic conduction starts. For the material compositions and dopants used for the waveguide layers in this work, typical values for the activation energies are 26.4 meV for the acceptor levels (C) and 6 meV for the donor levels (Si) [Ada93], so that all doping centers are activated at 300 K. Reducing the temperature to 200 K transfers the carrier concentration into the freeze-out range. Solving the neutrality condition at this temperature, the carrier concentration on the p-side is calculated to drop by $\sim 10\%$ (compared to room temperature) while full activation is still present on the n-side.

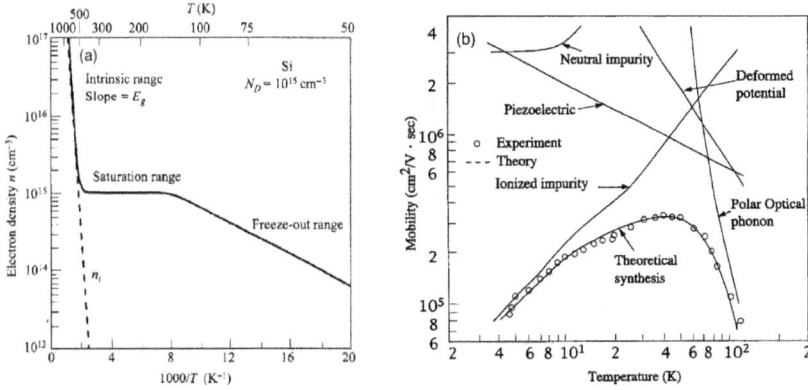

Figure 2.9: (a) Electron density plotted against the inverse temperature for a Si sample. Figure taken from [Sze01]. (b) Temperature dependence of carrier mobility, driven by scattering on ionized impurities and phonons. Figure taken from [Wol70].

The temperature dependence of bulk mobility is schematically drawn in figure 2.9 (b). Free carriers within the semiconductor are scattered by phonons as well as ionized impurities. While a reduction of temperature initially freezes out the phonons and the resulting longer mean free path of the carriers leads to an increase in mobility following $T^{-3/2}$, the scattering from ionized impurities increases with $T^{3/2}$. With the help of Matthiessen's rule, the combined mobility can be calculated. In the temperature range assessed, electron and the critical (as lower) hole mobility increase more than tenfold going from 300 K to 200 K, as the effect of the freezing out of phonons dominates. According to [Sze01], the reduced phonon scattering at lower temperatures further leads to a higher thermal conductivity, e.g. a lower thermal resistance.

The carrier concentration, mobility and resulting conductivity have previously been studied at the FBH on $Al_xGa_{1-x}As$ samples of bulk semiconductor that were epitaxially grown [Sin12]. Doping levels and composition of the semiconductor were chosen to reflect the epitaxial compounds used in the designs in later chapters. For all configurations which are to be used in the low doped waveguides, conductivity was shown to increase for decreasing temperatures in the range 200...300 K.

Chapter 3

Measurement Setup

The experimental setups for characterizing diode lasers are described in this chapter. Beginning with the standard, existing measurement stations at the FBH, the different procedures to determine the performance and check the quality of single emitters and diode laser bars at room temperature are outlined. An available setup for operating single emitters at temperatures down to $\sim 260\,\mathrm{K}$ is used to conduct first lower temperature measurements and serves as a blueprint for designing low temperature measurement setups. The development of a first low temperature station enables the characterization of bars up to currents of $2\,\mathrm{kA}$ and down to temperatures of $\sim 220\,\mathrm{K}$. A further optimization of this setup focuses on achieving temperatures of $\sim 200\,\mathrm{K}$ and improving the reliability of the system. For all low temperature measurement setups the challenge for the most accurate power calibration is thoroughly discussed going through multiple iterations of calibration routines and estimating the accuracy of the determined values. In the end, an accurate protocol is developed to confidently quote power levels of high power single emitters and bars, driven in quasi-continous wave mode up to highest powers at low temperatures.

3.1 Room temperature measurement setups

The FBH has developed a good reputation in developing diode lasers for a wide spectrum of applications. The final part of the development process is the quality check of realized designs by characterizing the fabricated diode lasers. For that purpose, multiple measurement stations are available at the institute to determine the performance of assembled as well as unassembled devices. Here, the stations to assess fabricated single emitters and bars for their electro-optical behavior around room temperature are described. First, the continuous wave (CW) and quasi-continuous wave (QCW) measurements are outlined, followed by the characterization of the spectrum. Next, an imaging system is introduced, which helps to determine the power distribution among the emitters on a single bar. Finally, a prototype setup for operating single emitters at temperatures below the freezing point of water is presented, which is used as a model for designing a low temperature measurement station.

3.1.1 Electro-optical and spectral characterization

The first step of characterizing high power diode lasers is to determine their optical output power at different drive currents while measuring the applied bias voltage. Fabricated single emitters (SE) mounted on screening submounts (SSM) are clamped into a measurement bracket which offers good thermal connection as well as access points for current carrying wires and temperature monitoring (figure 3.1).

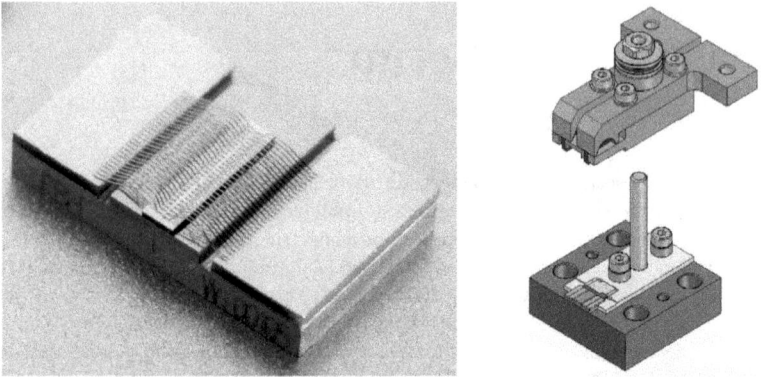

Figure 3.1: Left: single emitter with $L = 3\,\mathrm{mm}$ on screening submount, connected to two bond pads via gold bond wires. Right: sketch of measurement bracket for clamping of screening submounts.

On the left of figure 3.1, a SE mounted on a SSM is displayed, with its top side connected with gold bond wires to the two bond pads of the SSM. The SSM is positioned with the front facet facing outwards in the notch of the measurement bracket, depicted on the right and pressed firmly against the temperature sensor in the back of the notch. A temperature dependent platinum resistor is used, which has a nominal resistance of $R = 100\,\Omega$ at a temperature of $T = 0\,^\circ\mathrm{C}$ (PT100). It monitors the chip temperature and is used to ensure a constant operating temperature of the diode. The bracket's lid (green part in figure 3.1, right) presses down on the two bond pads and establishes an electrical connection to the pads via the electrically separated L-shaped Cu-blocks. Once inside the measurement bracket, the SE can be operated on and moved to different measurement stations by clamping down the bracket to a heatsink without altering the position of the SSM within.

The setup to determine the electro-optical characteristics for a SE is shown in figure 3.2. The measurement bracket holding the SSM is screwed onto a Cu-heatsink, which also serves an electrical connection to the bottom of the diode laser. The temperature of the Cu-heatsink is adjusted with a Peltier cooler, which dumps the extracted heat into a water cooled cycle. The Peltier cooler is electrically operated by a temperature controller (Newport Model 3150) which adjusts the power of the cooler for a constant operating temperature within a $\pm\,0.2\,\mathrm{K}$ window, measured with the PT100 pressed to the back of the SSM.

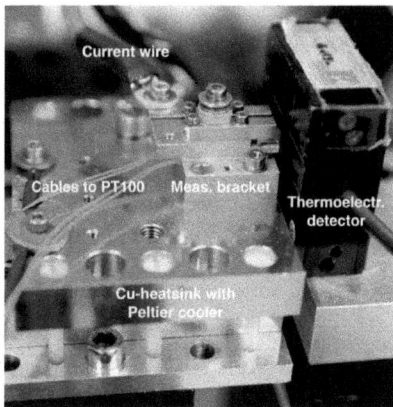

Figure 3.2: Setup for power measurement at room temperature with a thermoelectric detector, positioned in front of a measurement bracket with a SSM.

The current carrying wires are connected to one of the L-shaped blocks of the measurement bracket and to the Cu-heatsink. The second L-shaped block is used as an access point to measure the voltage with a multimeter (Keithley Model 2001) in 4-terminal configuration with an accuracy of $\sigma_U = \pm$ 5E-6 V, thus omitting the in-line resistance from bond wires or connections on the top side of the diode laser. The bottom side is connected by massive Cu-blocks (from the measurement bracket and the heatsink) and thus add negligible in line resistance. A current source (ILX 36085-12) supplies CW or QCW drive currents up to 120 A with an accuracy of $\sigma_I = \pm$ 0.02 A with pulse lengths of 100 μs or higher.
Directly in front of the measurement bracket, a thermoelectric detector (Gentec UP19K-15S-5W) is placed collecting the emitted light from the front facet and measuring the optical power with an accuracy of $\sigma_{P,therm.el.} = \pm$ 2.5% up to powers of 15 W. The detector is insensitive to wavelength shifts and averages the incident power over time. Thus, it is not feasible to use this detector for measuring the power of QCW operated diode lasers. The calibration procedure of the crucial power meter is described in [Cru13], and the detector is checked annually against international standards (ISO 11554). Due to the accurate readings of the current supply and the multimeter, the accuracy of the calculated efficiency from a CW measurement is carried by the accuracy of the thermoelectric detector and is given as ± 2.5%.
In the case of operating the diode in pulsed condition, a Si photodiode is used, which is connected to a port of an integrating sphere of 4-inch diameter. The emitted light is collected via an input port and dispersed within the sphere. Multiple optical density filters of different strengths are available to avoid saturation of the photodiode and achieve a high signal to noise ratio. The pulsed power reading of the photodiode is amplified via a transimpedance amplifier (Newport 1830c) and is recorded by an oscilloscope (Tektronix TDS-7104). The power signal is determined at the end of the pulse in a region where a stable plateau is reached by averaging over a gated region.

In order to get a power value, though, a wavelength sensitive calibration factor of the photodiode needs to be applied. This factor is found by measuring the CW power output of the diode with the thermoelectric detector and with the photodiode and comparing the slope of the power-current curve just above threshold. However, the additional waste heat in CW operation (compared to QCW operation) will lead to a slight red shift in wavelength and thus the CW calibration factor has to be applied with a margin of error. As the photodiode is very wavelength sensitive, this margin is estimated as $\sigma_{\lambda PhDiod} = \pm 3\%$. All in all, the measurement of QCW power is much less accurate than the CW power, as calculated from the addition of the standard deviations

$$\sigma_{P,QCW} = \sigma_{P,therm.el.} + \sigma_{P,PhDiod,CW} + \sigma_{\lambda PhDiod} \approx \pm 6\% \quad (3.1)$$

with $\sigma_{P,PhDiod,CW}$ the accuracy of the CW power signal of $\leq 1\%$ recorded by the photodiode. The QCW power accuracy translates directly into the QCW efficiencies, which is thus given as $\pm 6\%$.

The same measurement setup with some alterations is used to characterize mounted bars. Temperature is now measured and fixed on the Cu-heatsink by a PT100 sensor. Due to the mounting scheme of the bar, a 4-terminal configuration of measuring voltage across the diode is no longer possible and package resistance is always included in bar measurements. A different current supply (Beratron Berillium) is used to drive the bars up to higher currents of 400 A with an accuracy of $\pm 2\%$. The bars are only measured in QCW mode in the same way as the SE and the wavelength sensitive calibration factor gained from CW measurements of SE is used to gain the pulsed power.

The spectrum of the emitted light is analyzed in a spectrometer (OceanOptics HR4000) which is connected to a port on the integrating sphere by a fibre. The spectrometer covers a spectral range of 890 nm to 1050 nm with a resolution of ± 120 pm (FWHM) in full dynamic range. The gathered signal is integrated over time and recorded for each applied current. By integrating over the signal $\int S(\lambda)d\lambda$, the central wavelength as well as the spectral width $\Delta\lambda_{95\%}$ is found.

3.1.2 Histogram of diode laser bars

The power distribution across a fabricated QCW laser bar is checked with an imaging system. Using the same heatsink as for the electro-optical characterization, the bar is operated at a constant current and temperature. The emitted light from all emitters is imaged through two lenses and passes through a slit, which cuts off all but one emitter. On the other side of the slit, the resulting optical signal is recorded by a photodiode attached to an integrating sphere as depicted in figure 3.3.

The light emitted from the bar first passes through a fast axis cylindric collimating lens (focal length $f = 10$ mm) before being collected by a slow axis cylindric lens ($f = 25.4$ mm). At the position of the slit, the emitters are well separated and a single emitter can be isolated by adjusting the slit's aperture and the bar's lateral positioning via a movable stage the heatsink is attached to. The QCW power of this single emitter is collected by an integrating sphere and the power signal recorded by the attached photodiode connected to the transimpedance amplifier and oscilloscope as described in section 3.1.1. However, this time, the absolute power is not determined by applying a photodiode calibration factor. Instead, the power signals of all the emitters are recorded first by moving the bar laterally and thus imaging the adjacent emitter through the slit into the sphere. The sum of all power readings is now equal to the recorded power of the bar at that specific current and the distribution of power amongst the emitters is known.

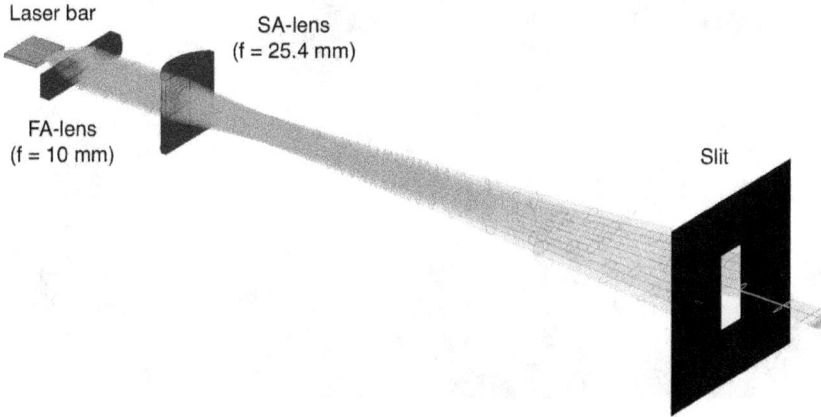

Figure 3.3: Imaging setup for determining the power distribution across a laser bar.

3.1.3 Low temperature prototype setup

The challenge of lowering the operation temperature of the diode lasers to sub-room temperatures lies in the accumulation of water (and at lower temperatures ice) on the surfaces at temperatures below the dew point. This can lead to short circuits and the damaging of the laser's facet. Thus, the laser diodes cannot be operated in the normal laboratory's atmosphere but within a space with a lower dew point.

Initially, a measurement setup was available which allowed the electro-optical characterization of SE at operating temperatures below the freezing point of water. The accumulation of dew is avoided by screwing the measurement bracket onto a heatsink placed within a single-walled chamber as depicted in figure 3.4.

As for the room temperature measurement station, the temperature of the heatsink in the chamber is regulated by a Peltier cooler, which is water cooled and electrically operated by a temperature controller. The heatsink is placed in front of an antireflection (AR) coated window, which is angled by 4° off the perpendicular plane to the beam axis, as to avoid back reflections into the diode laser. The measurement bracket with the SSM within is connected electrically as described in section 3.1.1 and temperature is monitored at the back of the chip again. All cables are fed through the chamber walls and are connected to the same devices used for the room temperature measurement. Once all connections are established, the chamber is closed with a lid and pressure within the chamber is reduced to 900 mbar.

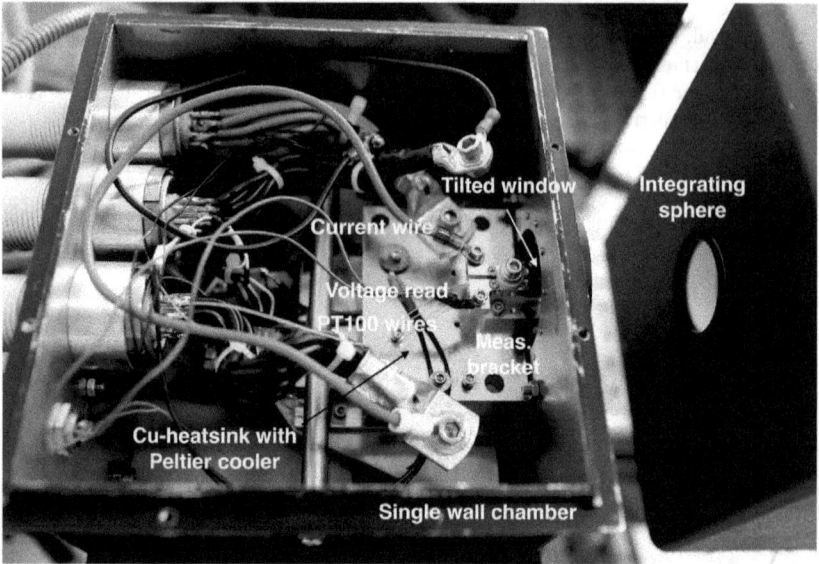

Figure 3.4: Available prototype setup for electro-optical characterization of single emitters down to operating temperatures of 258 K.

The emitted light of the diode is collected in an integrating sphere (positioned on the right side of figure 3.4) and analyzed following the same procedures used for QCW power measurements at room temperature. The calibration factor for the QCW power determination is found by comparing the slope of CW measurements recorded at room temperature with the thermoelectric detector at the standard measurement setup and a CW measurement in the vacuum chamber recorded with the photodiode. Lowering the operation temperature, however, shifts the wavelength of the emitted light. Taking a wavelength shift with temperature of ~ 0.35nm/K, the difference in wavelength between room temperature and the lowest temperature in the chamber of 258 K is 14 nm. Due to the wavelength sensitive photodiode, this change in wavelength will lead to a major difference in calibration factor. Furthermore, the wavelength sensitivity of the whole system of the AR coated window, the lining of the integrating sphere, the ND filters in front of the photodiode and the photodiode itself has to be taken into account. In order to track the wavelength dependence of the calibration factor, multiple different SE at wavelengths between 900 nm and 980 nm are measured in CW mode at room temperature on the standard measurement station and in the chamber. The resulting calibration factors at the different wavelengths are fitted with a polynomial and a calibration function $F(\lambda)$ is established. However, due to the limited number of SE with different wavelengths, this function is assumed to have an uncertainty of $\pm$ 5%, which translates directly into an uncertainty for the determined power of 5%.

3.2 Low temperature measurement setup: first generation

Based on the schematics of the prototype low temperature measurement station, the development of a first low temperature measurement setup able to reach temperatures of $\sim 220\,\mathrm{K}$ is described. This setup allows the electro-optical characterization of single emitters and bars over a wide power and temperature range. The system is able to drive single bars up to a current of $2\,\mathrm{kA}$ in QCW mode while tracking the pulse shape of current, voltage and power simultaneously. By establishing a new power measurement setup, the accuracy of power determination is enhanced and its wavelength sensitivity drastically reduced.

3.2.1 Physical setup of low temperature measurement station

The concept of operating the diode lasers in a sealed environment to avoid condensation at temperatures below the dew point of the laboratory atmosphere is adopted in the setup of the low temperature measurement station. The requirements for the measurement station are expanded to reach lower operation temperatures, enable higher thermal loads and facilitate the characterization of single emitters as well as bars. Since diode lasers will be measured over a wider range of temperatures and thus over a broader spectral range, a wavelength insensitive power measurement is required. Furthermore, possible damages to the devices at the highest powers necessitate the collection of all wanted data (such as electro-optical *and* spectral values) in one measurement run.

Similar to the prototype, a vacuum chamber is designed to house the heatsink which the measurement bracket of the SE or bars will be attached to (figure 3.5). Again, this heatsink is placed before a 3-inch round hole in the chamber wall, which is sealed by an AR-coated, tilted window, through which the optical power will be extracted.

The two main differences to the prototype setup lie in the temperature control of the heatsink and in the connection to supply currents up to $2\,\mathrm{kA}$ to diode laser bars. As the temperature difference between the heatsink and a possible cooling water cycle is $> 60\,\mathrm{K}$, no single stage Peltier cooler can be used. Multiple stage Peltier coolers offer the possibility to remove heat across a higher temperature difference, but the poor efficiency of the elements stack and huge amounts of waste heat just from the Peltier coolers will have to be extracted on top of the laser's waste heat. Instead, an external chiller (Huber CC-820w) is chosen to circle Silicone oil through a heatsink, thermally connected but electrically separated to the heatsink the lasers are clamped to. The Silicone oil remains liquid down to temperatures of $\sim 180\,\mathrm{K}$ and the chiller regulates the oil's temperature to keep a stable laser operation temperature, as measured by the PT100 temperature sensor in the chamber (either pressed to the chip for SE or to the heatsink for bars). Most of the cooling capacity is used to cool down the system, especially the massive chamber. Thus, the available capacity to extract the waste heat is limited and depends on the operation temperature. At a heatsink temperature of $\sim 225\,\mathrm{K}$ for example, the cooler is able to extract up to $\sim 50\,\mathrm{W}$ waste heat from the diode, and cooling capacity decreases rapidly for lower temperatures.

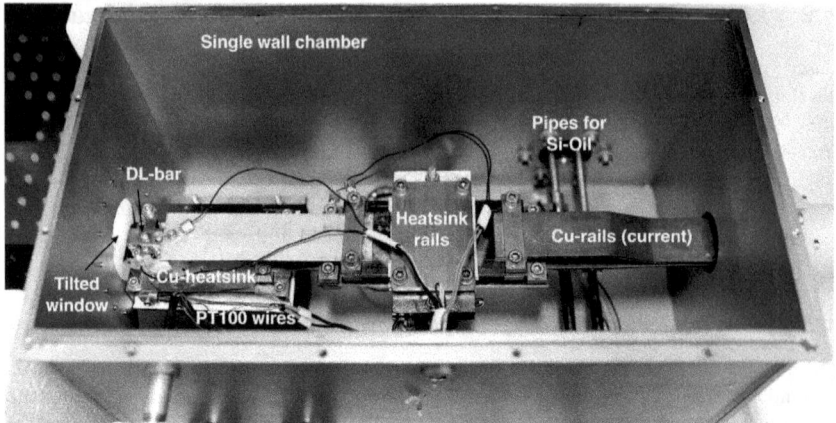

Figure 3.5: Picture of the physical setup of the measurement setup for single emitters and bars for temperatures down to $\sim 220\,\text{K}$.

The electrical connection for the laser bars is established by 3 mm thick, 3 cm wide Cu-rails, which are able to conduct high QCW currents up to 2 kA and duty cycles of up to 4%. The rails are fed through the chambers walls and can be connected to different current supplies. Within the chamber, they are cooled by a second separate heatsink as to not conduct the ambient heat directly to the laser diode. Three different current supplies are available depending on the drive conditions and current supply needed:

- ILX 3660 for SE measurements, supplying up to 20 A in CW and QCW mode

- Amtron CS412 P for high current SE measurements, supplying up to 50 A in QCW mode

- Amtron CS411 P for high current bar measurements, supplying up to 2 kA in QCW mode with duty cycles up to 4%, pulse widths 100-2000 µs, accuracy ± 1%

In case of the Amtron current supply being used to drive QCW laser bars, an in-line resistor of $R = 200 \pm 2\,\text{m}\Omega$ is introduced between the end of the current rails and the current supply. The voltage drop across the resistor is tracked by an Oscilloscope (Tektronix TDS5034B) and is used to determine the applied current with an accuracy of ± 1%.

Once the diode is clamped to the heatsink and the electrical connections have been made, the chamber's lid is closed and a low pressure of 880 mbar is established. The voltage for SE operated in CW mode is measured in a 4-terminal configuration with a multimeter (HP Model 34401A) with an accuracy of ± 8.5E-5 V. Voltage measurement during QCW operation is also performed in 4-terminal configuration (though the measured bias of bars includes the package resistance) and the pulsed signal is recorded by the oscilloscope with an 8 bit vertical resolution.

The light extracted from the window is collected in a 5.3-inch integrating sphere. Attached to the sphere are a photodiode, a fibre port and a thermoelectric detector (Gentec XLPF12-3S-H2) able to record average powers up 3 W with an accuracy of ± 2.5%. The Si-photodiode is connected to a transimpedance amplifier (Newport 1830c), which records CW power signals and amplifies the QCW signal which is tracked by the oscilloscope. A fibre connects the fibre port to a spectrometer (OceanOptics HR 4000) which covers a spectral range of 600 nm to 1030 nm with a resolution of ± 120 pm (FWHM) in full dynamic range. The intensity of the light reaching the fibre and the photodiode can be separately adjusted by introducing ND filters. While the use of a standalone thermoelectric detector would be beneficial to collect absolute (averaged) power values, the distance of the detector outside to the diode within the chamber requires too large an aperture of the detector, especially for smaller output powers of single emitters. Furthermore, the small distance to the chamber's cold wall would subject the thermoelectric detector to broadband radiation, depending on the operating temperature of the laser and thus lead to an under- or overestimation of recorded power. Additionally, no spectral or pulse form information can be gained simultaneously and a second measurement would have to be performed, at which point the device might already be damaged. This is especially critical for high power measurement. The chosen measurement approach of attaching a thermoelectric detector to an integrating sphere circumvents these problems, but introduces the question of power calibration for both CW and QCW measurements.

3.2.2 Power calibration

CW and QCW optical power signals are recorded by the photodiode as well as the thermoelectric detector, both attached to the integrating sphere. As diodes of different wavelengths are measured over a wide range of temperatures, a power calibration to translate the recorded signals into power values needs to be wavelength insensitive. The sensitivity of the photodiode depends strongly on the incident wavelength and thus, its absolute power reading is neglected (though the pulse shape of the emitted optical power is tracked on the oscilloscope).

The typical power levels recorded by the thermoelectric detector can be estimated by assuming that the collected power is dispersed homogeneously across the inner surface of the sphere (which is not true due to the large 2.5-inch input port and the other various ports not adding to the dispersion of light). The 1-inch port of the thermoelectric detector is thus illuminated by 0.9% of the light. The thermoelectric sensor has a low, but not negligible wavelength sensitivity and in combination with the AR coated window and the inner coating of the integrating sphere, a wavelength dependent calibration factor $F(\lambda)$ needs to be deduced. By comparing the absolute CW power of SE of various wavelengths (measured at room temperature at the calibrated standard measurement station in section 3.1.1) to the recorded CW power of the thermoelectric detector in the low temperature station, $F(\lambda)$ is determined as displayed in figure 3.6.

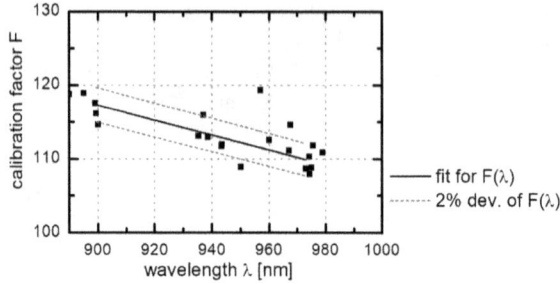

Figure 3.6: Power calibration factor for the low temperature setup, as determined from measuring SE of different wavelengths at two measurement stations and comparing the extracted slope.

Instead of comparing the power at a specific current between the two measurements setups, which can lead to high uncertainties, the slope of the power-current curve shortly above threshold is compared. The absolute values for F shown in figure 3.6 match the estimated fraction of light reaching the thermoelectric detector on the integrating sphere ($1/0.9\% \approx 110$) and the weak wavelength dependence of the detector is visible (6% difference in F for a 80 nm λ-difference). The calibration factor can be confidently stated with an accuracy of $\pm$ 2%. The accuracy of the power measurement at the low temperature measurement station can thus be estimated by adding the accuracies of the thermoelectric detectors at the standard and the low temperature station and the calibration factor, resulting in $\sigma_P = \pm$ 7%.

Similarly, a wavelength dependent calibration factor is determined for QCW measurements at the low temperature station with fixed pulse conditions of pulse length $\tau = 1.2$ ms and frequency $f = 10$ Hz by comparing the slope to the CW slope at the standard measurement station. This is based on the assumption, that shortly above threshold, CW and QCW powers coincide. The specific pulse condition is chosen to match the requirements for the optical pumps of Yb:YAG based solid state laser systems.

The restriction to two operation modes (CW and one specific QCW configuration) proves to be problematic when measuring devices at different duty cycles to limit their thermal load, as done frequently during high power bar measurements. While it is possible to match the low current slope of a 1.2 ms, 10 Hz measurement to the slope of a measurement with different QCW conditions in the same current regime, problems occur when selecting shorter pulse widths ($\tau < 300$ µs). Here, the rise time of the current signal is no longer negligible compared to the pulse width and strong deviations from a rectangular pulse shape appear. Furthermore, the Amtron 2 kA source switches between different current stages within for different current ranges, which leads to an abrupt change in rise time and thus current pulse shape. This translates directly into the pulse shape of the power, and the averaged power over the pulse changes significantly at the switch point. The exact shape of the optical power in the flanks of the pulse can not be accurately traced with the measurement setup, as the rise time of the power pulse is superimposed by a longer time constant of the photodiode / transimpedance amplifier combination. The rise time of the power does not match the rise time of current and thus does not correctly represent the optical power pulse reaching the thermoelectric detector.

This change of pulse shape is visible in power kinks in the power-voltage curve. Measuring bars, kinks in the curve could also result from emitter failures. Thus, matching the slope of the 1.2 ms, 10 Hz measurement to that of measurements with shorter pulse widths can lead to an incorrect power estimation.

3.3 Final Cryo-measurement setup

The challenges seen in the first low temperature measurement setup regarding lower operating temperatures as well as a universal power calibration are addressed by design changes to the physical setup. A better insulation of the chamber and the use of a more powerful cooler enable the operation of single emitters and bars down to temperatures of ~ 200 K. Furthermore, the power measurement setup is revised which makes it possible to track the pulse shape of the emitted QCW power accurately. This allows for a precise determination of power independent of drive conditions or pulse length while still using a wavelength insensitive detector. A new power calibration based on the pulse shape analysis is established which allows the declaration of in-pulse power, peak power of the pulse and average power within a gated time interval.

3.3.1 Changes to the physical setup

A second measurement chamber is designed which focuses on using the chiller's cooling power on the laser's heatsink rather than on the chamber itself. This is achieved by three alterations: while leaving the dimensions of the chamber untouched, the walls are lined with multiple plastic sheets of 1 mm thickness to reduce convection cooling from the heatsink. Second, the thickness of the copper rails supplying the current pulses is reduced to 1 mm to reduce the heat flow across them to the outside. Within the chamber, the heatsink for the rails is omitted and the rails are connected directly to the laser's heatsink. Lastly, the chiller itself is upgraded to a more powerful model (Huber CC-906w). All these changes now enable operation temperatures of ~ 200 K, the targeted temperature of cryogenically cooled pumps for solid state laser systems, with sufficient heat extraction capacity for diode laser bars operated with low duty cycles.

The challenge of determining the power of QCW measurements with shorter pulse lengths is solved by replacing the combination of the photodiode / transimpedance amplifier by a fast InGaAs photodiode (Thorlabs PDA20CS-EC), which has an attached amplifier. Connecting this photodiode to the oscilloscope, the rise times of the power signal match the rise times of the current signal. Kinks in the power-current curve can now be undoubtedly assigned to either an emitter failure or a pulse shape change (except for the unlikely case that both happen at the same current step). The precise depiction of pulse shape now also allows a more accurate power calibration for QCW measurements of all pulse lengths.

3.3.2 QCW power calibration by pulse shape analysis

For QCW measurements of SE, the pulse shapes of the bias voltage and the power are recorded together with the spectrum and the averaged power signal of the thermoelectric detector at each current value. For QCW bars, the pulse shape of the voltage drop across the inline resistor is additionally stored, allowing a detailed analysis of the current pulse. With the detailed knowledge of its pulse shape, the integrated power signal P_{avg} of the thermoelectric detector calculated by the recorded average power and the calibration factor $F(\lambda)$ (see figure 3.6) can be transferred into absolute pulse power values. The calculations necessary are subsequently laid out, following [Fre16].

The repeating pulses and the pulse shape are analyzed to gain the pulse frequency f and width τ, respectively. For an ideal rectangular pulse shape, the in-pulse power P_{pulse} is calculated via

$$P_{\mathrm{pulse}} = \frac{P_{\mathrm{avg}}}{f \cdot \tau} \tag{3.2}$$

As described previously, the actual pulse shape is never perfectly rectangular, as the pulse's flanks are affected by rise and fall times (especially pronounced for short pulses and at power switches within the current supply) and drops in power signal over the course of the pulse are occasionally observed (for example due to self heating). Thus, P_{pulse} does not sufficiently describe the pulsed power. Rather, the optical power at a certain point of the pulse (for example P_{max}) or averaged over a certain part of the pulse (P_{gate}) are sought. The power signal can be converted into an absolute power value by integrating the photodiode signal $S_{\mathrm{pd}}(t)$ over the duration of the pulse (see figure 3.7), which is proportional to P_{pulse}

$$P_{\mathrm{pulse}} = \frac{A}{\tau} \times \int_{t_0}^{t_1} S_{\mathrm{pd}}(t)\mathrm{d}t \tag{3.3}$$

with A as a proportionality factor, t_0 and t_1 the pulse beginning and end, respectively.

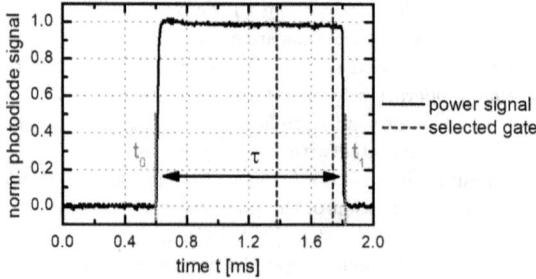

Figure 3.7: Optical pulse with pulse length $\tau = 1.2\,\mathrm{ms}$, measured by the InGaAs photodiode, with pulse beginning and end t_0 and t_1, respectively. A gated region of 0.65 and 0.95 τ is displayed (dashed), in which the average power is determined.

The proportionality factor A can be determined by combining equations 3.2 and 3.3:

$$A = \left(\frac{P_{\mathrm{avg}}}{f}\right) / \left(\int_{t_0}^{t_1} S_{\mathrm{pd}}(t)dt\right).$$

A is now calculated at each single current and pulse shape changes are accounted for by the integration of the signal. The power $P(t)$ at each time of the pulse is thus given as

$$P(t) = A \cdot S_{\mathrm{pd}}(t). \tag{3.4}$$

It is now possible to state the values for the average power over the pulse (P_{pulse}), the maximum power within the pulse (P_{max}) or the average power within a certain part of the pulse (P_{gate}). The values for the current, voltage and efficiency must be given accordingly for the same chosen case. For SE measurements, the current pulse shape is not recorded (no voltage drop over an in-line resistor) but is assumed to have a rectangular shape so that $P_{\mathrm{pulse}} = P_{\mathrm{max}} = P_{\mathrm{gate}}$. Unless stated otherwise, the values given in this work for QCW measurements of bars and SE represent the power, current, voltage and efficiency averaged over a gate between 0.65 τ and 0.95 τ at which point the power has reached a stable value.

An overview of the different measurement stations is given in table 3.1 and the specifications regarding test conditions are summarized.

Table 3.1: Existing and developed measurement setups and their specifications.

Setup	Param.	T [K]	Laser	mode	$I_{\mathrm{max.}}$ [A]	$P_{\mathrm{avg.,max.}}$ [W]	τ [ms]
RT setup	P, U, I, λ	288 - 358	SE	CW	120	15	
				QCW	120	not def.	0.04- 1
			bar	QCW	500	not def.	0.05 - 5
imag. syst.	pow. distr.	288 - 358	bar	QCW	500	not def.	0.05 - 5
low temp. prot.	P, U, I, λ	258 - 318	SE	CW	20	not def.	
				QCW	20	not def.	0.01 - 1000
first low temp.	P, U, I, λ	218 - 298	SE	CW	20	300	
				QCW	50	300	100 - 3000
			bar	QCW	2000	300	0.1 - 4
final low temp.	P, U, I, λ	200 - 298	SE	CW	20	300	
				QCW	50	300	100 - 3000
			bar	QCW	2000	300	0.1 - 4

Chapter 4

Benchmark Designs (Iteration 0)

In chapter 4 the behavior of established efficient, high power vertical FBH designs at decreasing temperatures is analyzed. The initial iteration aims at identifying the prerequisites of a vertical structure which are necessary for high power and high efficiency operation at low temperatures. Furthermore, the impact of the lateral layout on the performance at low temperatures is explored. First, available structures designed for room temperature operation are assessed in single emitter configuration and the structure dependent benefit of lower operating temperatures is tracked. Based on their performance two vertical designs are identified as promising candidates for high power, high efficiency operation at low temperatures. Subsequently, these benchmark designs are processed into single emitter and bars with two different lateral layouts and lengths. Measurement results of the devices are presented at room temperature as well as at temperatures close to the targeted 200 K. At the lowest temperature, bars reach peak powers of 1.45 kW with a power conversion efficiency of 30%. The limitations to higher efficiencies at low temperatures are discussed and the reduction of series resistance is singled out as the most important factor.

4.1 Performance of established vertical designs

Numerous vertical structures have been developed at the FBH for various applications, most of which are designed to operate at room temperatures. Iteration 0 aims at identifying the structures capable of reaching the required power levels of 20 W from a 100 µm wide single emitter (equivalent to 1.5 kW from a 1 cm wide high fill factor bar). The effects of the design of the active region, the thickness and composition of the waveguide as well as the asymmetry of the vertical structure on the efficiency at these high powers are assessed on single emitter levels. Also, the benefit of decreased operation temperatures on the power and efficiency is studied.

4.1.1 High power room temperature optimized designs

Selected vertical designs differ in asymmetry, doping, thickness and composition of the waveguide (WG) and layout of the active region. The asymmetry is given by the thickness of the p-WG compared to the total WG thickness (in percent). To illustrate the parameters pertaining to the design of the WG, the general vertical layout is displayed in figure 4.1, showing the thickness d as well as the n- and p- WG. The displayed exemplary structure is roughly symmetric (47%) as n- and p-WG are similar in thickness.

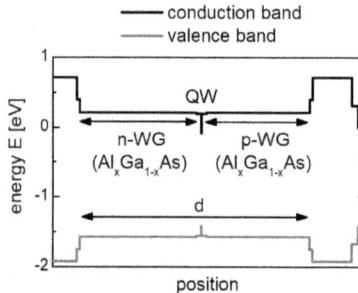

Figure 4.1: Conduction and valence band of an exemplary vertical design, showing the division of total WG thickness d into p- and n-WG.

A total of seven designs was chosen which can be grouped into three types of structures:

- Asloc configuration: thick WG (large optical cavity) in which the active region is asymmetrically located (thinner p- than n-WG). No significant changes to AlGaAs-WG compositions (Al-content at $x = 15\%$) and room temperature wavelength $\lambda = 975\,\mathrm{nm}$ between the structures.

  - Asloc1: 2.4 µm thick WG with 0.95 µm p-WG (40% asymmetry) and a single quantum well (SQW).
  - Asloc2: 3.9 µm thick WG with 1.7 µm p-WG (44% asymmetry) and a SQW.
  - Asloc3: 4.8 µm thick WG with 1.7 µm p-WG (35% asymmetry) and a triple quantum well (TQW).

  Goal: explore effect of asymmetry and total WG thickness on performance.

- Asloc configuration with fixed high asymmetry (17%), doping and thin total WG thickness, but changing Al-content in the WG and n-cladding. Wavelength at room temperature is 940 nm.

  - AlVar1: 1.4 µm thick WG with 250 nm p-WG and a SQW; aluminum content in the WG of 20%.
  - AlVar2: 1.4 µm thick WG with 250 nm p-WG and a SQW; aluminum content in the WG of 25%.

Goal: test influence of WG composition on performance.

- Edasloc configuration: extreme double asymmetric. Large WG, extremely asymmetric (6%) [Has14], also asymmetry in thickness of cladding (thick n-clad). The Al-content in the waveguides is changed as well as the active zone layout.

  – Edas1: 2.65 μm thick WG with 150 nm p-WG; Al-content in the WG of 25%; thick InGaAs QW surrounded by thick GaAsP barriers.

  – Edas2: 2.65 μm thick WG with 150 nm p-WG; Al-content in the WG of 30%; QW as well as GaAsP barrier thickness reduced.

Goal: investigate extreme case of vanishing p-WG, the effect of Al-content in the waveguide and the impact of the active zone layout.

Assembled single emitters featuring the different vertical designs were available at the start of iteration 0. The vertical designs had been processed into $w = $ 90-100 μm wide broad area emitters with different lengths.

4.1.2 Performance of thick waveguide Asloc configuration at decreasing temperature

At the beginning of iteration 0, a measurement station supporting temperatures down to $T_{HS} \sim 258$ K was available to assess the impact of decreased operation temperature on diode laser performance. The measurement procedure is given in chapter 3 for the prototype low temperature measurement setup.

The properties of the selected $L = 4$ mm long diodes with the Asloc design are summarized in table 4.1.

Table 4.1: Properties of tested $L = 4$ mm diodes using the Asloc configuration with $x = 15\%$ Al-content in the WG.

Design	d_{WG} [μm]	QW	w [μm]	mount.	R_f [%]	R_r [%]	asym. [%]	dop. p-WG [cm^{-3}]	dop. n-WG [cm^{-3}]
Asloc1	2.4	SQW	100	p-up	1	95	40	6E16	6E16
Asloc2	3.9	SQW	90	p-down	5	98	44	1E17	5E16
Asloc3	4.8	TQW	90	p-down	1	98	35	1E17	5E16

The difference in R_f among the diodes will lead to a change in power emitted from the front facet due the difference in α_m. The higher number of QWs in the Asloc3 configuration is expected to lead to a higher threshold current, as more carriers are needed in the active zone to reach transparency.

Three lasers per type were measured under quasi-continuous wave (QCW) condition with a pulse length $\tau = 1.2$ ms and a frequency $f = 10$ Hz in the temperature range $T_{HS} = 258\ldots298$ K. By choosing a low duty cycle pulse measurement, the higher thermal resistance of the p-up mounted diodes is not assumed to cause major deteriorations in power. For each vertical Asloc design, the diode reaching the highest optical output power was selected and its performance at $T_{HS} = 298$ K and 258 K is shown in figure 4.2.

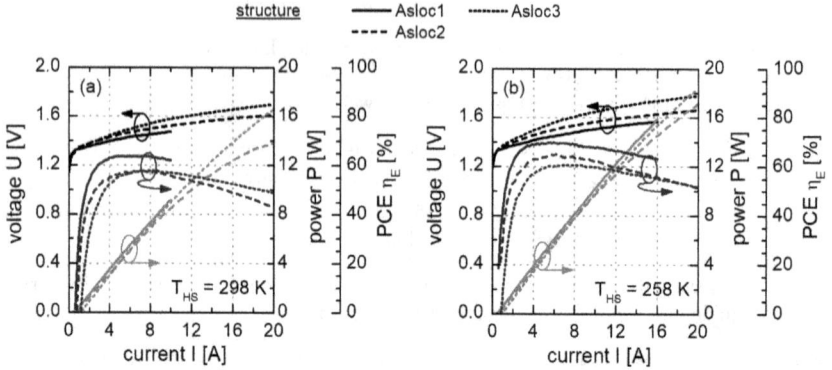

Figure 4.2: Voltage, power and efficiency of single emitters with the three Asloc designs at (a) T_{HS} = 298 K and (b) T_{HS} = 258 K, measured with 1.2 ms, 10 Hz pulses.

At room temperature, the Asloc1 design has the highest efficiency (figure 4.2 (a)) with peak efficiency being η_{max} = 64%. This is due to the lowest threshold current, the highest slope and lowest series resistance R_S (visible by the slope of the UI-curve) of the three Asloc structures. The diodes with the Asloc1 design were not measured up to the highest currents, as their facets were not passivated and thus risk of catastrophic optical mirror damage was high. The second most efficient structure features the Asloc3 design. Slope is equally high as for the Asloc1 design, but threshold current is visibly increased due to the higher number of quantum wells, leading to a much reduced peak efficiency. The high slope is sustained up to $I \sim 10$ A at which point slight roll over of the power is observed. The diode reaches a power of 16.5 W at the maximum current. However, the Asloc3 design has the highest series resistance seen by the higher slope of the voltage current curve which leads to a higher voltage at the high currents. Finally the Asloc2 design shows a similar threshold current to the Asloc1 design, but has the lowest slope which starts rolling over visibly at $I \sim 8$ A, which leads to an output power of 13.8 W at 20 A. The low slope is caused by the lower mirror loss due to the high R_f. Though the series resistance is lower than in the Asloc3 design, the strong roll over in power limits the efficiency strongly at high currents. The three vertical designs all show a non-linear voltage-current characteristics, as the UI-curve is bending and local values for the differential series resistance are reduced at high bias. This bending does not change the order of ranking the structures by their series resistance, and lower R_S designs have lower voltages at all currents. However, the reduction of series resistance at high currents limits the voltage and thus benefits the efficiency.

Across all designs, power and efficiency are much improved at T_{HS} = 258 K (figure 4.2 (b)). Still, the Asloc1 design performs best reaching peak efficiency of η_{max} = 69% due to its low threshold current, high slope and low series resistance. Roll over is much less pronounced at this temperature, only starting at $I \sim 10$ A. The Asloc3 design, suffering from the high series resistance and the high threshold current, sustains a constant slope up to the $\sim 12\,\mathrm{A}$, at which point it starts rolling over, reaching $18.3\,\mathrm{W}$ at the maximum current which represents a 11% power increase from the room temperature measurement. The increased threshold current (compared to the Asloc1 design) significantly reduces peak efficiency, though, in turn also limiting the efficiency at high currents. The design Asloc2 again experiences strong power roll over. Compared to the room temperature measurements, maximum output power is increased by 25% to $17.2\,\mathrm{W}$ at $20\,\mathrm{A}$, though. While peak efficiency is higher than the Asloc3 design due to the lower I_{th}, the deterioration in slope at high currents significantly reduces the efficiency. Even the lower series resistance compared to the Asloc3 design does not compensate for the roll over in power and efficiency at high currents between the two designs is the same.

All three designs with the Asloc configuration reach an output power of $16\,\mathrm{W}$ at the lower temperature and values at that point are summarized in table 4.2.

Table 4.2: Measured values at $16\,\mathrm{W}$ and peak efficiency from the diodes using the Asloc configuration at $258\,\mathrm{K}$.

Design	$I(16\,\mathrm{W})$ [A]	$U(16\,\mathrm{W})$ [V]	$\eta_{\mathrm{E}}(16\,\mathrm{W})$ [%]	$\eta_{\mathrm{E,max.}}$ [%]
Asloc1	16	1.57	63	70
Asloc2	18.2	1.65	53	65
Asloc3	16.9	1.74	54	60

Looking at table 4.2, the following conclusions can be generally drawn from the comparison between the Asloc configurations:

- A higher number of quantum wells increases threshold current, limiting the peak efficiency and thus also the maximum achievable efficiency at high currents. At the same time, more quantum wells seem to guarantee a more stable operation at high currents, as roll over in power is reduced.

- Series resistance is higher in structures using thick waveguides.

For a more thorough quantitive analysis of the behavior at lower temperatures, the characteristic parameters threshold current I_{th}, external differential efficiency η_{d} and series resistance R_{S} are extracted from the PUI-curves and subsequently analyzed. These values are obtained using a linear fit to the power current and voltage current characteristics in the interval $[I_{\mathrm{th}} + 0.2\,\mathrm{A}, I_{\mathrm{th}} + 1.6\,\mathrm{A}]$. With the knowledge of the central wavelength λ_{c}, η_{d} can be calculated from the slope $S = \mathrm{d}P/\mathrm{d}I$ using equation 2.14. The external differential efficiency is better suited for comparing various structures which might lase at different wavelengths, as it accounts for the wavelength dependence of the slope. Furthermore, the widening of the band gap with decreasing temperature also causes a shift in wavelength [Ada93] which is excluded by using η_{d}.

For the three Asloc configurations, the characteristic parameters are plotted against the heatsink temperature in figure 4.3.

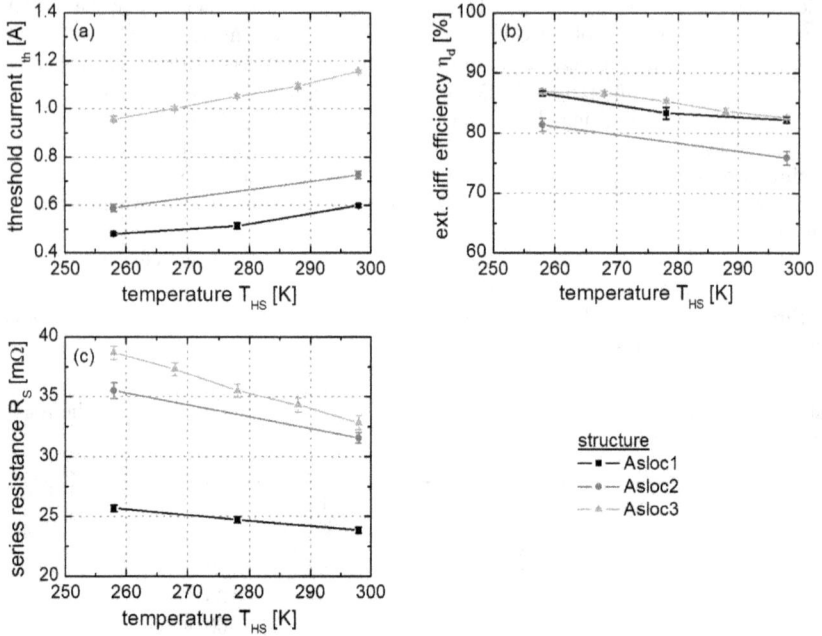

Figure 4.3: Characteristic parameters versus T_{HS} for the three Asloc designs extracted from linear fits. (a) Threshold current; (b) external differential efficiency; (c) series resistance.

In figure 4.3 (a), the development of threshold current with heatsink temperature is displayed. As already seen from the power current characteristics, the Alsoc3 design has the highest I_{th} of the three designs, almost twice as high as the Alsoc1 design, as the structure uses a TQW (compared to a SQW in the other designs). All designs show a reduction of threshold current with decreasing temperature. In the case of the Asloc1 design, threshold current is lowered from $I_{th} = 0.6\,A$ at $T_{HS} = 298\,K$ to $0.48\,A$ at $258\,K$. For all three designs, I_{th} drops by $\sim 20\%$ in this temperature span, highlighting the benefit of reduced temperature on this characteristic parameter.

The temperature dependence of the external differential efficiency is shown in figure 4.3 (b). While the designs Asloc1 and Asloc3 show a similar η_d, external differential efficiency is much lower for the Asloc2 with the higher R_f, i.e. lower α_m. At $T_{HS} = 298\,K$, $\eta_d = 82\%$ for the Asloc1 design and 76% for the Asloc2 design. With decreasing temperature, η_d rises to 87% and 81% for the Asloc1 and Asloc2 design, respectively. Since any increase in external differential efficiency is directly coupled to the increase in power conversion efficiency, the advantage of lower operation temperatures is obvious.

Finally, the series resistance is plotted against the heatsink temperature in figure 4.3 (c). As stated before, the series resistance is higher for the structures with thicker waveguides.

For example, at T_{HS} = 298 K series resistance is 24 mΩ for the Asloc1 design, with R_S = 31 mΩ and 33 mΩ for the Asloc2 and Asloc3 design, respectively. With decreasing temperature, the series resistance is increased. However, the change in series resistance is not the same for all structures. Between T_{HS} = 298 K and 258 K, R_S increases by 2 mΩ for the Asloc1 design, but by 4 mΩ for the Asloc2 and even 6 mΩ for the Asloc3 design. Thus, the thickness of the WG does not only influence the absolute series resistance but also its temperature dependence. Especially the thickness of the p-WG has a prominent effect on the series resistance. This can be seen by comparing the Asloc2 to the Asloc3 design. While the WG is 23% thicker for the Asloc3 design, the p-WG has the same thickness. The 40% thicker n-side only increases R_S by 2 mΩ at room temperature. In contrast, comparing the Asloc1 and Asloc2 designs, the WG is 62% thicker with the p-WG being 79% thicker for the Asloc2 structure. While the increase in n-WG thickness is comparable (50% thicker), the series resistance is now significantly increased by 7 mΩ for the Asloc2 design. Thus, a thin p-WG is beneficial for lower absolute R_S as well as lower increase in R_S with temperature.

The measurements of the designs using the Asloc configuration reveal the negative impact of thicker WG on the series resistance. Not only is the absolute resistance higher in thicker waveguide, but R_S also increases more at lower operation temperatures. Especially the p-waveguide is crucial for low series resistance, highlighting the benefit of an asymmetric waveguide structure. While the use of multiple QW reduces power saturation at high currents, it significantly increases threshold current and thus limits peak efficiency. The highest powers and efficiencies are expected from the Asloc1 design, which uses a SQW and a thin overall waveguide.

4.1.3 Performance of Asloc configuration with changing waveguide composition at decreasing temperature

The properties of the available L = 4 mm diodes with the AlVar designs are summarized in table 4.3.

Table 4.3: Properties of tested L = 4 mm, w = 90 µm diodes using the AlVar designs with a SQW.

Design	d_{WG} [µm]	mount.	R_f [%]	R_r [%]	asym. [%]	Al_{WG} [%]	dop. p-WG [cm^{-3}]	dop. n-WG [cm^{-3}]
AlVar1	1.4	p-down	2	95	17	**20**	1E17	5E16
AlVar2	1.4	p-down	2	95	17	**24**	1E17	5E16

The emitter had passivated facets for high power operation and wavelength was centered around 940 nm at room temperature. The diodes were tested in QCW mode (1.2 ms, 10 Hz) in the temperature range T_{HS} = 258...298 K using the same measurement setup as for the Asloc diodes. The performance at the highest and lowest tested heatsink temperature is shown in figure 4.4.

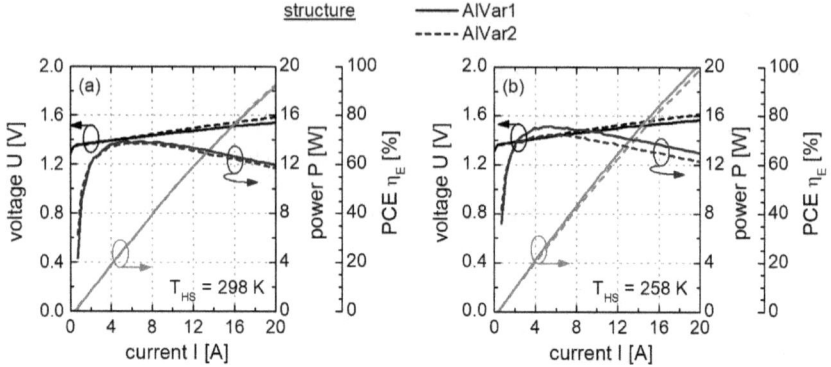

Figure 4.4: Voltage, power and efficiency of single emitters with the two AlVar designs at (a) T_{HS} = 298 K and (b) T_{HS} = 258 K.

At T_{HS} = 298 K, both AlVar structures display a very similar behavior (figure 4.4 (a)). Threshold current and slope are the same and only at $I > 16$ A the two power current characteristics can be distinguished, as the AlVar1 diode experiences stronger roll over. Both designs experience power saturation above ∼ 12 A as the PI curves start rolling over. There is, however, a visible difference in the series resistance between the two designs, as the voltage current characteristic of the AlVar1 design has a lower slope than the AlVar2 structure. Both designs show little bending of the UI curve and series resistance stays constant up to the highest powers. The lower series in the AlVar1 designs leads to a slightly higher efficiency at higher powers, with η_{max} = 69% for both designs. At $P = 18$ W, efficiency is 59% and 60% for the AlVar1 and AlVar2 design, respectively.

The differences in the electro-optical characteristic of the two designs are much more pronounced at T_{HS} = 258 K (figure 4.4 (b)). Here, the AlVar1 design shows a higher slope, reaching an output power of $P = 20$ W with an efficiency of 65%. The AlVar2 design has a lower slope as well as a higher series resistance both of which cause a significantly lower efficiency compared to the AlVar1 design. As at room temperature, the voltage-current characteristic stays linear up to the highest currents. Peak efficiency is 76% and 72% for the AlVar1 and AlVar2 design, respectively, and at $P = 18$ W, efficiency is 67% and 63%. All in all, the lower operation temperature leads to a major increase in output power and efficiency, mainly due to an increased slope.

For comparison with the Asloc designs, the performance of the two AlVar designs at a power of 16 W and a temperature of 258 K is summarized in table 4.4.

Table 4.4: Measured values at 16 W and peak efficiency from the AlVar designs at 258 K.

Design	$I(16\,\text{W})$ [A]	$U(16\,\text{W})$ [V]	$\eta_E(16\,\text{W})$ [%]	$\eta_{E,max.}$ [%]
AlVar1	15.1	1.53	69	76
AlVar2	15.7	1.57	65	72

Following the analysis of the characteristic parameters for the three Asloc configurations, I_{th}, η_d and R_S are extracted to provide a more detailed comparison between the two AlVar structures. From fits to the power current and voltage current characteristics in the interval $[I_{th} + 0.2\,A, I_{th} + 1.6\,A]$, e.g. just above threshold, the characteristic parameters are deduced at each temperature for the two AlVar designs. In figure 4.5, their temperature dependence is plotted.

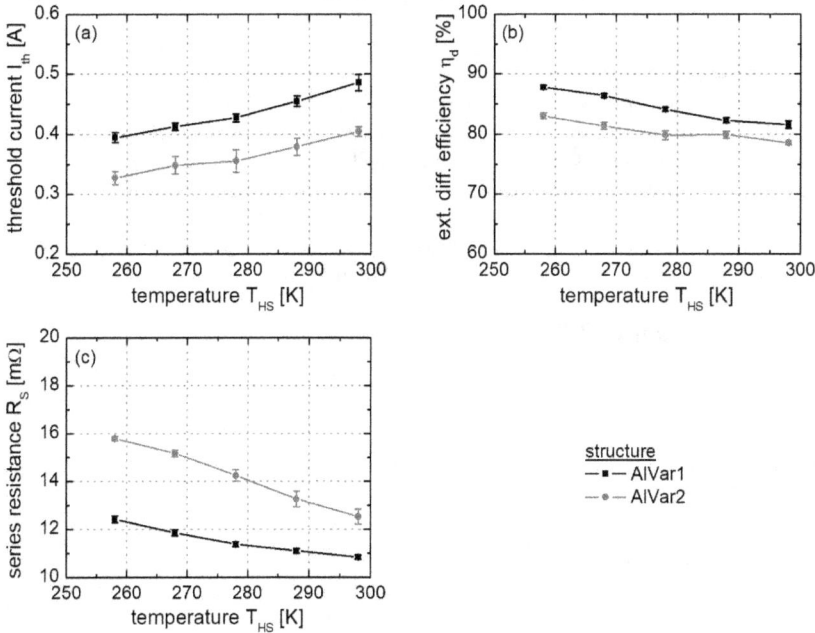

Figure 4.5: Characteristic parameters versus heatsink temperature for the two AlVar designs extracted from linear fits. (a) Threshold current; (b) external differential efficiency; (c) series resistance.

The dependence of threshold current on the heatsink temperature for the two AlVar designs is shown in figure 4.5 (a). At all temperatures, the AlVar1 design has a higher threshold current than the AlVar2 design and both designs experience a decrease in I_{th} with decreasing temperature. There is, however, a stronger reduction for the AlVar1 design with the lower Al-content in the waveguide. Here, threshold current decreases from $I_{th} = 490\,mA$ at $T_{HS} = 298\,K$ to $390\,mA$ at $258\,K$, compared to a change from $400\,mA$ to $330\,mA$ for the AlVar2 design at the respective temperatures.

In figure 4.5 (b) the development of external differential efficiency with heatsink temperature is traced. The AlVar1 design shows a higher η_d than the AlVar2 design at all temperatures. While external differential efficiency increases with decreasing T_{HS} for both structures, the AlVar1 design benefits from a stronger improvement of η_d. The value for

45

η_d increases from 79% to 83% for the AlVar2 design when going from room temperature to $T_{HS} = 258$ K. Simultaneously the AlVar1 design's external differential efficiency rises from 82% to 88% in the same temperature regime.

The temperature dependence of the series resistance for the two AlVar designs is plotted in figure 4.5 (c). The lower Al-content in the waveguide leads to a lower R_S for the AlVar1 design at all temperatures. With decreasing T_{HS}, the series resistance increases for both structures. However, R_S rises faster with lower temperature for the AlVar2 design. From 12.5 mΩ at room temperature, the series resistance grows to 15.8 mΩ at $T_{HS} = 258$ K for the AlVar2 design, while R_S only increases from 10.8 mΩ to 12.4 mΩ for the AlVar1 design in the same temperature interval.

For the Al-contents used, a lower Al-content benefits the performance at low temperatures. While threshold current is higher, the reduced series resistance and higher external differential efficiency are crucial for high powers and high efficiencies at high powers. Furthermore, the development of the characteristic parameters with decreasing temperature for the lower Al-content structure is advantageous, as threshold current decreases faster, η_d increases quicker and series resistance grows slower. This is especially important as the final targeted operation temperature is well below the lowest assessed temperature here.

4.1.4 Performance of Edas configuration at decreasing temperature

In table 4.5, the properties of the 4 mm long diodes using the Edasloc structures are listed.

Table 4.5: Properties of tested $L = 4$ mm, $w = 100$ µm diodes using the Edasloc structures with a SQW.

Design	d_{WG} [µm]	mount.	R_f [%]	R_r [%]	asym. [%]	Al$_{WG}$ [%]	dop. p-WG [cm^{-3}]	dop. n-WG [cm^{-3}]
Edas1	2.65	p-up	5	96	6	25	1E17	6E16
Edas2	2.65	p-up	5	96	6	30	1E17	6E16

The two main differences between the vertical structure pertain to the Al-content in the waveguide and the layout of the active region (not listed in table 4.5). The Edas1 design used a thicker InGaAs QW with thicker surrounding barriers of a different material composition than the Edas2 design. The facets of all diodes of both designs had not been passivated. Together with the p-up assembly onto CuW heatspreaders this led to a lower maximum operation current in the measurements as not to risk catastrophic optical mirror damage. The diodes were measured in the previously used measurement setup in QCW mode (1.2 ms, 10 Hz) in the temperature range $T_{HS} = 258\ldots298$ K. In figure 4.6 the electro-optical characteristics at $T_{HS} = 298$ K and 258 K are shown.

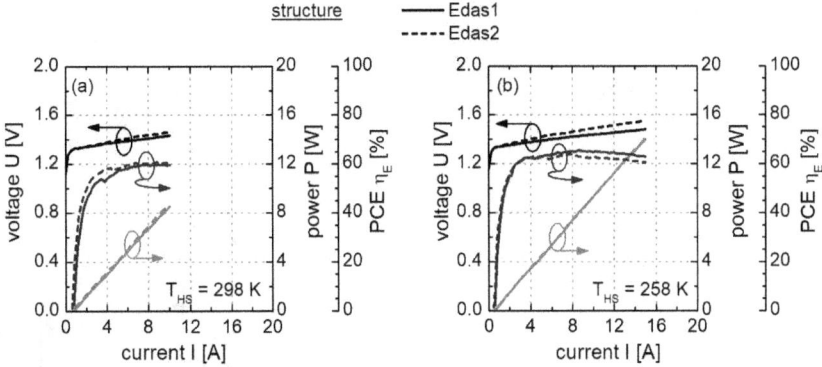

Figure 4.6: Voltage, power and efficiency of single emitters with the two Edasloc designs at (a) $T_{HS} = 298\,K$ and (b) $T_{HS} = 258\,K$.

The performance of the two Edasloc designs at $T_{HS} = 298\,K$ is displayed in figure 4.6 (a). Up to the tested current of 10 A, the two structures show a similar behavior. Threshold current is slightly lower for the Edas2 design, while no significant difference in slope can be seen. Both structures also show no signs of roll over in the power current characteristics to the maximum current. Series resistance is higher for the Edas2 design as can be seen from the slope of the UI-curve and is expected to deteriorate efficiency at higher currents stronger than the lower R_S of the Edas1 design. There is little deviation from linearity in the voltage-current characteristic thus R_S stays constant for the assessed currents. For both designs peak efficiency is 60% which is reached close to the maximum current.

At $T_{HS} = 258\,K$, however, the difference between the two Edasloc designs (figure 4.6 (b)) is visible which had been tested up to $I = 15\,A$ at this temperature. The difference does not stem from the power current characteristics as threshold current is only slightly smaller for the Edas2 design and slope is the same within error so that both graphs lie on top of each other. The voltage power curves diverge however, as the Edas2 design shows a significantly higher series resistance. Still, the UI-curve is very linear and little bending is observed. The higher R_S of the Edas2 design deteriorates efficiency, especially at the maximum current. Peak efficiency is 66% and 64% for the Edas1 and Edas2 design, respectively, decreasing to 63% and 61% at the maximum power of $P = 14\,W$.

At the lowest temperature of 258 K, the Edas designs do not reach powers of 16 W. Instead, the measured values at a power of 14 W are listed in table 4.6.

Table 4.6: Measured values at 14 W and peak efficiency from the Edas designs at 258 K.

Design	$I(14\,W)$ [A]	$U(14\,W)$ [V]	$\eta_E(14\,W)$ [%]	$\eta_{E,\text{max.}}$ [%]
Edas1	15	1.48	63	66
Edas2	15.7	1.55	61	64

In order to gain more insight into the different behavior of the two Edasloc designs, the characteristic parameters are extracted from the measurements at the different temperatures by fitting the power current and voltage current characteristics in the interval $[I_{th} + 0.2\,\text{A}, I_{th} + 1.6\,\text{A}]$. The values for I_{th}, η_d and R_S are plotted in figure 4.7 against the temperature.

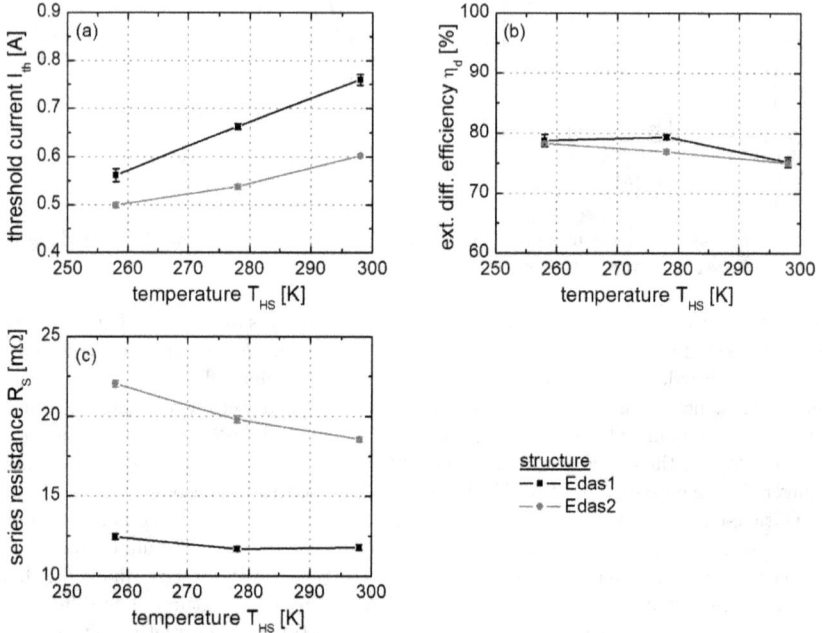

Figure 4.7: Characteristic parameters versus heatsink temperature for the two Edasloc designs extracted from linear fits. (a) Threshold current; (b) external differential efficiency; (c) series resistance.

In figure 4.7 (a), the temperature dependence of the threshold current is shown for the two Edasloc designs. The Edas1 structure has an increased I_{th} at all temperatures, compared to the Edas2 design, with threshold current decreasing for both designs with lower temperatures. Similar to the discussion of the AlVar designs, this increase is caused by the decrease in Al-content in the waveguide, though the difference in the active zone layout may also affect the threshold current and may be superimposed. As already seen from the discussion of the measurement results, the Edas1 design benefits from a more rapid decrease of threshold current with decreasing temperature, in line with the results in the previous section on AlVar designs. From $T_{HS} = 298\,\text{K}$ to $258\,\text{K}$, I_{th} is reduced from $760\,\text{mA}$ to $560\,\text{mA}$ for the Edas1 design, while the Edas2 design shows only a decline from $600\,\text{mA}$ to $500\,\text{mA}$ in this temperature range.

The external differential efficiency of the two Edasloc designs is plotted in figure 4.7 (b) against T_{HS}. Both structures have $\eta_d = 75\%$ at room temperature which increases to 78% for the Edas2 and 79% for the Edas1 structure at 258 K. The generally low η_d is due to the high front facet reflectivity of 5% which limits the slope in turn for a lower threshold current. However, the $\sim 3\%$ point increase in external differential efficiency is not sufficient to motivate a lower operation temperature. The reason for the small improvement in η_d at lower temperatures remains unknown and needs to be studied separately. For the purpose of identifying promising designs for low temperature operation, the little improved η_d rejects the use of Edasloc structures.

The temperature dependence of the series resistance of the two Edas structures is displayed in figure 4.7 (c). Here, a major difference can be seen between the designs, as the Edas1 structure has a significantly lower R_S at room temperature of 11.8 mΩ compared to the 18.6 mΩ of the Edas2 design. This advantage of the Edas1 design becomes more pronounced at lower temperatures, as series resistance grows faster for the Edas2 design. This behavior is attributed to the different Al-contents in the waveguide in accordance with the measurement results from the AlVar designs in the previous section. At $T_{HS} = 258$ K, $R_S = 12.5$ mΩ and 22 mΩ for the Edas1 and Edas2 design, respectively, meaning that at this temperature the Edas2 design has a 76% higher series resistance than the Edas1, compared to the 58% at room temperature.

The Edasloc structures do not show sufficient improvement at lower operation temperatures. This is mainly due to an only small increase in external differential efficiency, which is already low due to a high front facet reflectivity. At the same time, however, the series resistance is low and also changes little with temperature, especially for the Edas1 design. The low R_S together with a high linearity of the voltage-current curve, however, does not sufficiently benefit efficiency at high powers to compensate for the low and little improved η_d.

4.1.5 Selection of benchmark designs

After the performance evaluation of the available designs, the following observations are made for the assessed temperature regime:

- The number of quantum wells influences the magnitude of the threshold current and thus limits the achievable peak efficiency occurring at low currents. Also, high efficiencies at high currents can not be achieved with a low peak efficiency. At the same time, a higher number of quantum wells leads to lower roll over in output power at high currents.

- The thickness of the waveguide strongly affects the series resistance. Not only is the absolute series resistance bigger for designs using a thicker waveguide, but R_S also rises quicker with lower temperatures.

- The asymmetry of the waveguide also impacts the series resistance. A thin p-waveguide lowers R_S and causes a smaller increase at lower temperatures. Furthermore, the bending of the voltage-current curve is reduced and series resistance stays constant over a wider current range. In the most extreme case of Edasloc structures, however, the increase of external differential efficiency at low temperature is reduced.

49

- The composition of the waveguide, namely the Al-content, strongly affects the series resistance. Lower Al-contents lead to a lower as well as less temperature sensitive R_S. Simultaneously, the threshold current is negatively impacted by a reduced Al-content.

Comparing the structures, the Asloc designs offer a big improvement in output power and efficiency when lowering the temperature. Especially the Asloc1 design displays a strong reduction in threshold current together with a high external differential efficiency as well as the lowest series resistance of all the Asloc configurations. However, compared to the Asloc designs with the thin p-waveguide (AlVar1 and AlVar2) and the Edasloc structures, the series resistance is much higher. The designs AlVar1 and AlVar2 show the highest powers and efficiencies at all temperatures of all structures due to a low threshold current, high external differential efficiency and low series resistance. Here, the AlVar1 designs reaches higher efficiencies at high powers than the AlVar2, as series resistance is lower due to the lower Al-content in the waveguide. Finally, the Edasloc structures only exhibit a slight performance increase with lower temperature. While series resistance is very low and barely changes with temperature for the Edas1 design, the little increase in external differential efficiency at lower temperature limits higher powers and high efficiencies. By these observations, the structures Asloc1 and AlVar1 are chosen for a full laser process, in which the designs are to be fabricated into laser bars.

4.2 Low temperature performance of benchmark designs

As discussed in section 4.1, two vertical designs are processed into single emitters and laser bars. Single emitters of different lateral layouts are tested at decreasing temperatures and the efficiency limiting factors are assessed for their importance by analyzing the characteristic parameters. Furthermore, the effect of emitter width and cavity length on the characteristic parameters is studied. Bars are measured at two temperatures and their performance is compared to the measurement results of the single emitters. The most promising design is tested on single emitter level over an extended temperature range to gain insights into the behavior of the characteristic parameters. Also, bars with this design are tested up to high currents and achieve output powers far above 1 kW.

4.2.1 Fabrication of high power, high efficiency single emitters and bars

Two promising designs had been selected in section 4.1 for a full laser process based on their performance on single emitter level. The vertical designs Asloc1 and AlVar1 (from here on out referred to as Design1 and Design2, respectively) are grown using metal organic vapor phase epitaxy (MOVPE) on 3 inch substrates. Subsequently the wafers are processed into single emitter and bars with various lateral layouts. Single emitters are fabricated with stripe widths of 90 and 200 μm. High fill factor (FF) 1 cm wide bars, e.g. bars with a high occupancy of lasing area, are processed into a lateral design containing 18 emitters with a structured contact of $w = 400\,\mu\text{m}$, $L = 6\,\text{mm}$ (FF = 72%). Bars containing single emitters with cavity lengths $L = 4\,\text{cm}$ and 6 cm from Design1 and with $L = 6\,\text{cm}$ from Design2 were cleaved. Together with laser bars of both vertical designs containing high FF bars, their facets are passivated using ZnSe for high power operation [Res05], before being coated with $R_f = 1\%$ and $R_f = 98\%$ for the front and rear facet reflectivities, respectively. The single emitters are cleaved from the bars and mounted p-side down onto CuW heat spreaders using AuSn solder. The n-side is contacted with Au bond wires. The bars are first sandwiched between two CuW heatspreaders using AuSn and the sandwich is then soldered p-side down onto a conduction cooled package (CCP). The n-side contact was established by using a thin Cu-foil which was connected via PbSn solder to the top of the sandwich. An exemplary fully assembled single emitter as well as a bar on CCP is shown in figure 4.8.

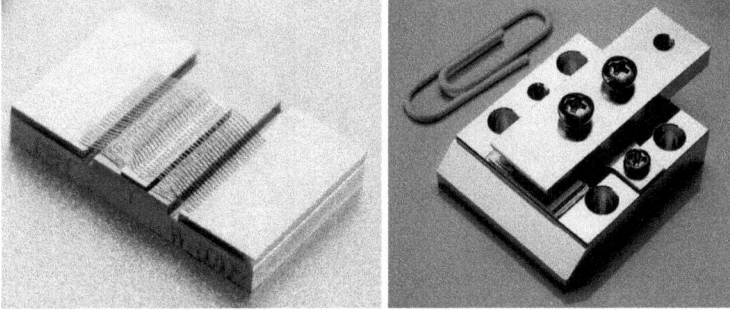

Figure 4.8: Mounted lasers in single emitter configuration on CuW heat spreader (left) and bar format, sandwiched between CuW heat spreaders and soldered onto a CCP (right).

4.2.2 Design1: Impact of lateral layout and length on performance

Single emitters with Design1 are assembled with two different stripe widths (90 µm and 200 µm) and lengths (4 mm and 6 mm) for a total of 4 configurations. The lasers are clamped to a heatsink and tested under QCW condition (1.2 ms, 10 Hz) in the temperature range $T_{HS} = 218 \ldots 298$ K using a newly acquired measurement setup described in chapter 3 (first low temperature setup). The electro-optical characteristics at $T_{HS} = 298$ K and 218 K are shown in figure 4.9. Here, the power, voltage and efficiency are plotted not against the total current I applied, but against a current i per stripe width w similar to [Wen00]:

$$i = I/w. \tag{4.1}$$

At the same time, the absolute power P also needs to be scaled to get the power p per stripe width:

$$p = P/w. \tag{4.2}$$

By using this current per stripe width and the power per stripe width, the 90 µm and 200 µm lasers can be more adequately compared. By design, we expect a wider emitter to sustain higher total current and reach higher output powers. This becomes very clear when looking at the configuration of a bar e.g. a 75% FF bar driven with $I = 1\,\text{kA}$. If the bar is made up from 75 emitters of 100 µm, each emitter is subjected to $I = 13.3\,\text{A}$, while each emitter in a bar consistent of 37 emitters of 200 µm width sees $I = 27\,\text{A}$. Overall, the current per contacted width on the bar is the same, though.

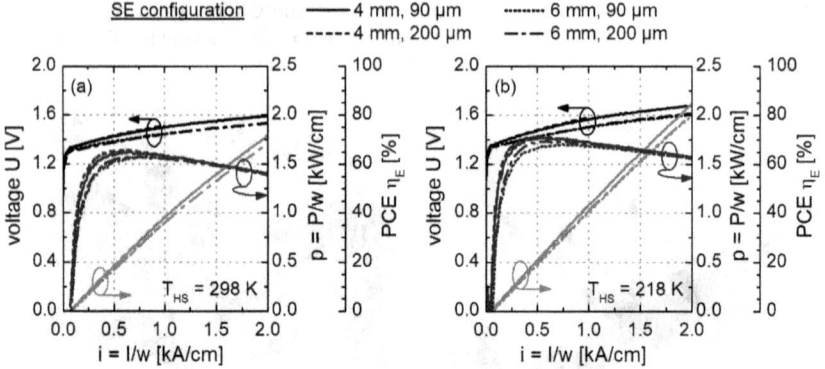

Figure 4.9: Power per stripe width p, voltage U and power conversion efficiency η_E versus current per stripe width i for the vertical Design1 and different single emitter layouts for (a) $T_\text{HS} = 298\,\text{K}$ and (b) $T_\text{HS} = 218\,\text{K}$.

In figure 4.9 (a), the electro-optical characteristics of 4 single emitter configurations at $T_\text{HS} = 298\,\text{K}$ is displayed. The main difference is seen between the $L = 4\,\text{mm}$ and 6 mm long emitters. The longer cavities show a higher threshold current compared to the 4 mm lasers as well as a lower slope. However, this can easily be explained by the higher mirror loss α_m of the short diodes using equations 2.5, 2.8 and 2.13. Since all diodes have the same facet coating, the longer resonators lead to a lower slope following equation 2.14. While the lower mirror loss of the longer diodes reduces threshold current density J_th according to equation 2.8, the bigger pumped area due to the longer cavities lead to a higher threshold current per stripe width.

While the differences in power-current characteristics are thus due to the mirror loss and can be fine tuned for a high power operation point by choosing adequate reflectivities, the long resonators offer a much reduced voltage which can be seen by the lower slope of the voltage-current per stripe curve. This stems from the bigger contact area of the longer stripes. The peak efficiency is reduced from $\eta_\text{max} = 66\%$ to 63% for the longer stripes due to the lower slope and the higher threshold current. However, at the maximum current per stripe width, the lower series resistance leads to an efficiency of 56% for all configurations. Small differences exist in comparing the different stripe widths of emitters with the same length. For both lengths, the 200 µm wide emitters show a slightly higher slope, which is also visible in the peak efficiency which differs between the 90 µm and 200 µm chips from 65% to 66% for the shorter and 62% to 63% for the longer cavity length, respectively.

At T_{HS} = 218 K, the electro-optical characteristics of the different single emitter configurations lead to similar conclusions (figure 4.9 (b)). Longer cavity lengths show a lower slope due to the reduced mirror loss as well as higher threshold currents per stripe width. At the same time, the $L = 4$ mm chips require a higher bias voltage, caused by the higher slope of the voltage-current per stripe curve. This means, that while the shorter cavities have a higher peak efficiency of $\eta_{max} = 71\%$ (compared to 69% for the $L = 6$ mm lasers), at the maximum current per stripe width of 2 kA/cm the reduced voltage of the longer stripes leads to the same efficiency of all diodes of 62%. Small differences are present between the different stripe widths, which can not be clearly identified from figure 4.9 but will have to be analyzed more thoroughly.

Comparing the performance at the two temperatures, the benefit of low temperature operation on efficiency and power output is evident. At the maximum current per stripe width, which corresponds to a current of $I = 1.5$ kA of a FF = 75% bar, the output power on bar level rises from $P = 1.28$ kW at room temperature to 1.52 kW at T_{HS} = 218 K for the 4 mm diodes, while power conversion efficiency rises from 56% to 62%. For maximum power, the $L = 4$ mm are the best choice, but at much higher currents per stripe width, the lower voltage of the $L = 6$ mm will eventually lead to higher efficiencies. The choice of bar layout thus needs to take into account the targeted maximum current as well as whether the focus is put on highest powers or efficiencies at high powers.

In order to explore the influence of the lateral layout in more depth and in line with the analysis of the different vertical structures in section 4.1, the characteristic parameters of the different stripe layouts are tracked over temperature by linearly fitting the power per stripe current per stripe as well as the voltage-current per stripe characteristics just above threshold in the interval $[I_{th} + 0.2$ A, $I_{th} + 1.6$ A$]$. This time, the threshold current and the series resistance are corrected by the emitter width and are thus replaced by the threshold current per stripe width I_{th}/w and by the resistance times stripe width $R_S \times w$. The temperature dependence of the extracted characteristic parameters is displayed in figure 4.10.

In figure 4.10 (a), the threshold current per stripe width is shown against the heatsink temperature. The threshold current for all emitter configurations decreases strongly in the assessed temperature range. At all temperatures, the lower threshold current at shorter cavity lengths is clearly visible. For the $L = 4$ mm diodes, the wider stripe width benefits from a reduced threshold current per stripe width by 5 A/cm. This is attributed to lateral current spreading which is present at the edges of the contact [Wen00]. While its magnitude should not be dependent on stripe width, the relative contribution to threshold current is increased for the narrower stripes.

The external differential efficiency is plotted against T_{HS} in figure 4.10 (b) for the four emitter configurations. In all cases, η_d increases with decreasing temperature, reaching 90% at the lowest temperature for the design $L = 4$ mm, $w = 200$ μm. As already seen from the electro-optical characteristics, the longer resonators have a lower external differential efficiency due to the lower mirror loss. However, this representation shows the benefit of wider stripes more clearly. For a given length, the $w = 200$ μm emitters have a 2% points higher η_d than the 90 μm diodes, which is again attributed to the lower contribution of lateral current spreading to a reduction in internal differential efficiency [Wen00]. The external differential efficiencies at T_{HS} = 258 K match the previously determined values for the Asloc1 design in section 4.1.2 at the lowest assessed temperature.

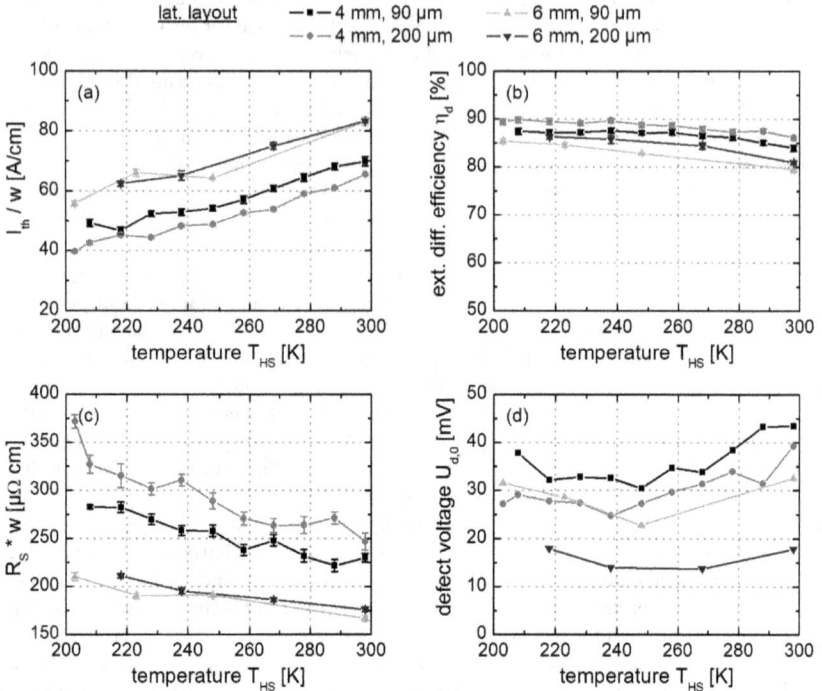

Figure 4.10: Temperature dependence of characteristic parameters of various single emitter layouts using vertical Design1; (a) Threshold current per stripe width I_{th}/w, (b) external differential efficiency η_d, (c) series resistance times stripe width $R_S * w$ and (d) defect voltage $U_{d,0}$.

The series resistance times stripe width is traced against the heatsink temperature in figure 4.10 (c). The benefit of longer resonators is explicit in the much reduced series resistance times stripe width at all temperatures compared to the $L = 4\,mm$ diodes. With decreasing temperature, $R_s * w$ increases for all emitter configurations. However, the longer emitters experience a smaller increase in series resistance per stripe width. For example, $R_S * w$ for the $L = 6\,mm$, $w = 90\,\mu m$ diode increases from $170\,\mu\Omega cm$ at $T_{HS} = 298\,K$ to $190\,\mu\Omega cm$ at $223\,K$ while for the $L = 4\,mm$, $w = 90\,\mu m$ emitter it increases from $230\,\mu\Omega cm$ at room temperature to $280\,\mu\Omega cm$ at $T_{HS} = 218\,K$. Regarding the influence of stripe width on the series resistance times stripe width, the narrower stripes have a reduced $R_S * w$ which also increases less with decreasing temperature than the $w = 200\,\mu m$ diodes. This effect is most visible for the $L = 4\,mm$ diodes. A possible explanation for this behavior is again the lateral current spreading which plays a bigger role in narrower stripes and increases the contact area. The increase in bulk layer conductivity at lower temperatures amplifies the effect of lateral current spreading, effectively increasing the area along the vertical current path. The intersection of the fit to the voltage-current per stripe width curve following equation 2.17, the defect voltage at $0\,A$ $U_{d,0}$, is plotted against the heat sink temperature in figure 4.10 (d). Except for the emitter with $L = 6\,mm$, $w = 200\,\mu m$ no major differences between the stripe widths and cavity lengths can be detected, with $U_{d,0}$ ranging between $44\,mV$ and $23\,mV$ over the whole temperature span. Even the stripe with the biggest contact area only has a minimum defect voltage of $14\,mV$ which does not change significantly with temperature. As the impact of $U_{d,0}$ on the voltage at high currents is very limited, and its magnitude is not critically affected by temperature, it will not be further taken into account in the optimization process of high efficiencies at high powers.

All in all, the assessment of emitter layout has shown the impact of stripe width and cavity length on the performance at high currents. First, while longer emitters benefit from significantly reduced series resistance, their threshold current is visibly increased and slope is reduced (for given facet reflectivities). Increasing mirror loss by reducing R_f will increase slope and output power at high currents, but further enlarge threshold current and thus reduce peak efficiency. Against these downsides of long resonators, the low series resistance benefits the efficiency only at high currents at which point peak efficiency will decrease less. Second, for the evaluated stripe widths, broader emitters have higher external differential efficiency as well as a lower threshold current. Simultaneously, the resistance of wider emitters is higher. These three effects are attributed to a higher importance of current spreading for narrower stripes.

Subsequently, the high FF bars featuring the vertical Design1 were tested at room temperature and at $T_{HS} = 223\,K$ under QCW condition ($1.2\,ms$, $10\,Hz$). The power current characteristics of an exemplary bar is shown in figure 4.11 which compares the performance of the bar to that of a $L = 6\,mm$, $w = 200\,\mu m$ single emitter scaled to bar level. For the comparison, the bars were only tested up to $I = 0.4\,kA$ as not to damage the diodes. In figure 4.11 (a), the output power, voltage and efficiency at $T_{HS} = 298\,K$ of a bar and a single emitter scaled to bar level are compared. Threshold current matches perfectly, but the bar suffers from a lower slope of $S = 0.98\,W/A$ than the single emitter ($1.03\,W/A$). This corresponds to a 5% decrease in slope, equivalent to one of the 18 emitters on the bar failing. At the same time, the voltage of the bar is much increased, as seen by a higher slope of the voltage current curve. The series resistance is increased from $0.24\,m\Omega$ for the scaled single emitter to $R_S = 0.67\,m\Omega$. This increase is attributed to the package resistance of the bar. While the single emitter is measured in a 4-terminal configuration, minimizing all in line resistances, the resistance of the n-side Cu-foil is included in the voltage measurement of the bar. Furthermore, the structured contact of the $400\,\mu m$ wide emitters in the bar reduces the actual contact width to $200\,\mu m$. Though the contact ridges are only $5\,\mu m$ apart and current spreading under the contact will eventually increase the width of the current path to the total $400\,\mu m$, the reduced contact area is also responsible for some of the increased voltage. The higher series resistance as well as the lower slope result in a significantly reduced efficiency of 52% at the maximum current compared to the 62% of the single emitter.

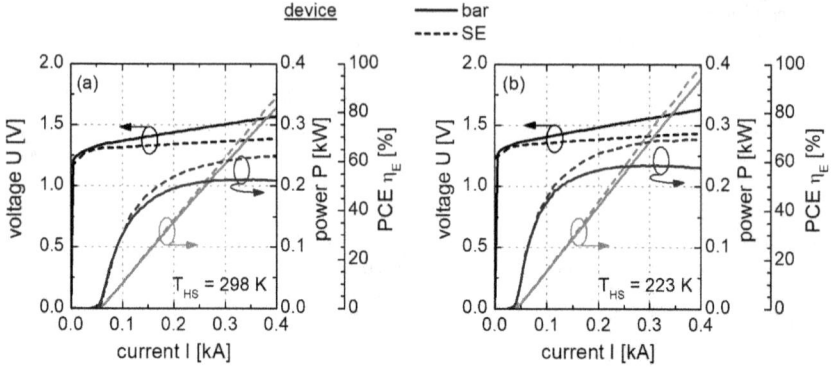

Figure 4.11: Comparison of power-, voltage- and efficiency-current characteristics between a bar (solid) and a single emitter (dashed; scaled by fill factor of the bar) at (a) T_{HS} = 298 K and (b) T_{HS} = 223 K

The performance at the lower temperature of T_{HS} = 223 K is compared for the bar and the single emitter in figure 4.11 (b). For the power current characteristics, the threshold current of the bar is reduced compared to the room temperature measurement and is as expected from the single emitter results. Slope is much increased to $S = 1.06\,\text{W/A}$ at the lower temperature and the bar reaches an output power of 0.38 kW at the maximum current. However, compared to the single emitter, slope is again reduced by 5%. This means, that the decrease in slope is not temperature dependent, as could be expected from additional mounting strain for the bar but is rather resulting from one emitter failure. The most significant difference between bar and scaled emitter is seen again in the voltage current characteristic. The higher voltage of the bar is caused by the higher series resistance of 0.76 mΩ at this lower temperature compared to 0.28 mΩ for the single emitter. Power conversion efficiency at the maximum current is with 58% much higher than at room temperature, but significantly reduced compared to the 69% for the single emitter.

In summary, the analysis of the lateral layout performed on fully processed devices using Design1 revealed the benefit of wider stripes. The relative contribution of current spreading is reduced in 200 µm wide stripes compared to 90 µm emitters, increasing external differential efficiency and reducing threshold current. At the same time, longer resonators provide low series resistance due to the increased contact area, while reducing external differential efficiency and increasing threshold current. Comparing bar and single emitter measurements the deteriorating effect on efficiency of package resistance in the bar assembly is discovered.

The characteristic parameter extracted from the high FF bar and SE measurements are summarized in table 4.7. SE results are scaled by the FF of the bar.

Table 4.7: Characteristic parameters of high FF bars and scaled values for the SE using the vertical structure Design1.

Device	I_{th} [A]	S [W/A]	R_S [mΩ]	$\eta_E(I_{max.})$ [%]
298 K				
Bar	58	0.98	0.67	52
SE	58	1.03	0.23	62
298 K				
Bar	43	1.06	0.76	58
SE	45	1.12	0.28	69

4.2.3 Design2: Performance up to high currents

Following the analysis of the influence of lateral emitter layout on the behavior at different temperatures, the performance of the vertical Design2 is assessed. In section 4.1, single emitters with the Design2 showed the highest powers and efficiencies due to high external differential efficiency, low threshold current and low series resistance. The single emitters with $L = 6\,\mathrm{mm}$, $w = 200\,\mu\mathrm{m}$ which are coated with $R_f = 1\%$ and $R_r = 98\%$ for front and back reflectivities, respectively, are first tested in continuous wave (CW) operation up to low currents for highest measurement accuracy at various temperatures $T_{HS} = 218 \ldots 338\,\mathrm{K}$ in 10 K steps. Exemplary power, voltage and efficiency versus current curves at three different temperatures are shown in figure 4.12.

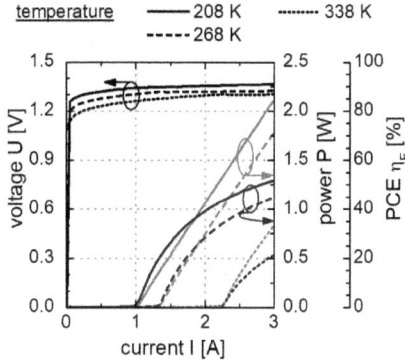

Figure 4.12: Performance of single emitter with the vertical Design2 at temperatures $T_{HS} = 218\,\mathrm{K}$, $268\,\mathrm{K}$ and $338\,\mathrm{K}$.

The power current characteristics at the three temperatures $T_{HS} = 218\,\mathrm{K}$, $268\,\mathrm{K}$ and $338\,\mathrm{K}$ reveal a strong rise in threshold current with higher temperatures. This increase is non-linear as threshold current grows by $\sim 30\%$ between $T_{HS} = 218\,\mathrm{K}$ and $268\,\mathrm{K}$, but by $\sim 100\%$ between 268 K and 338 K. Little difference is seen in the slope of the power current characteristic at the three temperatures so that the decrease in power at the maximum current is mainly resulting from the change in threshold current. As wavelength is shifted significantly at the different temperatures ($\lambda = 951$ nm at 218 K and 977 nm at 298 K), the constant slope shows that external differential efficiency is not strongly improved at lower temperatures.The voltage current characteristics show a decrease in voltage which mainly results from the shift in wavelength, e.g. the narrowing of the bandgap with higher temperatures. However, a difference in slope of the UI curve is also visible, as resistance at higher temperatures is decreased.

The characteristic parameters are analyzed to track the efficiency limiting factors over temperature. Similar to previous evaluations, the PUI curves are linearly fitted in the interval $[I_{th} + 0.25\,A, I_{th} + 1.25\,A]$ and the characteristic parameters are plotted against the heatsink temperature in figure 4.13.

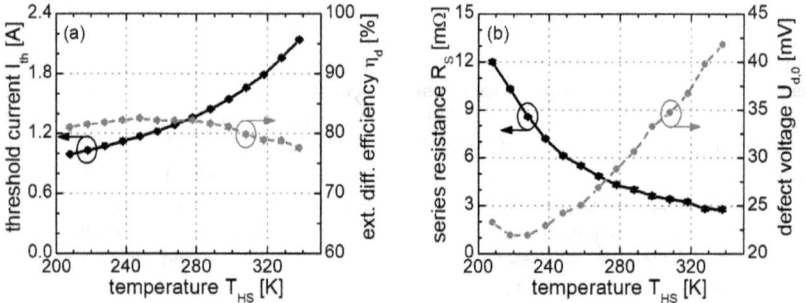

Figure 4.13: Characteristic parameters at different heatsink temperatures for Design2; (a) threshold current (squares) and external differential efficiency (circles); (b) series resistance (squares) and defect voltage at 0 V (circles).

In figure 4.13 (a) the temperature dependence of threshold current and external differential slope efficiency is traced. As seen from the PUI curves, the threshold current increases strongly with increasing temperature from 1 A at $T_{HS} = 218$ K to 2.1 A at 338 K, showing an exponential behavior as expected from equation 2.21. At the same time, the external differential efficiency does not show a significant improvement with decreasing temperature. From 78% at the highest temperature, η_d increases slightly to 83% at $T_{HS} = 248$ K before decreasing to 82% at the lowest temperature. Also, between room temperature and the targeted operating temperature, η_d shows no improvement. This is in contrast to the findings for the design AlVar1 in section 4.1.3, in which diodes reached $\eta_d = 88\%$ at $T_{HS} = 258$ K. Taking into account the different mirror loss (different facet coatings as well as different lengths) explains part of the discrepancy, though the newly processed diodes still fall 3% points short of the expected external differential efficiency. Compared to diodes of the same lateral layout using the vertical Design1, which reaches $\eta_d = 86\%$ at the lowest temperature, this is a significantly lower external differential efficiency. At room temperature, the two designs have an equally high value for η_d, but Design2 shows little to no improvement at lower temperatures.

The behavior of series resistance and defect voltage at $0\,A$ with T_{HS} is shown in figure 4.13 (b). Due to the large contact area of the $200\,\mu m$ wide, $6\,mm$ long emitter, series resistance is only $3\,m\Omega$ at $338\,K$ and increases to $10\,m\Omega$ at the lowest temperature. While the series resistance is significantly lower than for Design1 at room temperature, it increases faster with decreasing temperature and at $T_{HS} = 218\,K$ the two designs show the same R_S. In the assessed temperature range, $U_{d,0}$ falls strongly with temperature from $42\,mV$ to $23\,mV$. The beneficial impact on the efficiency is however negligible compared to the strong increase in R_S, especially at the high targeted currents.

From the low-current evaluation of the single emitters, a decrease in temperature does not improve the performance of Design2 significantly and Design1 shows a more promising behavior at low temperatures due to the higher external differential efficiency. For the measurements to high currents, the single emitter was tested in QCW condition (1 ms, 16 Hz) at room temperature and 223 K. Subsequently a bar with the Design2 and lateral layout of Motiv C was driven to highest currents in QCW mode (1.2 ms, 10 Hz) at room temperature and compared to the single emitter. The difference in duty cycle (1.6% for the single emitter, 1.2% for the bar) is due to the drive capacities of the different current supplies used and should not lead to significant deviations between the devices. After the test to high currents, the single emitters on the bar were imaged through a telescopic lens setup into an integrating sphere (chapter 3) and the power distribution across the bar was mapped at a current of $I = 0.16\,kA$ at 298 K at the same pulse conditions. The room temperature results of the bar and single emitter measurements are shown in figure 4.14.

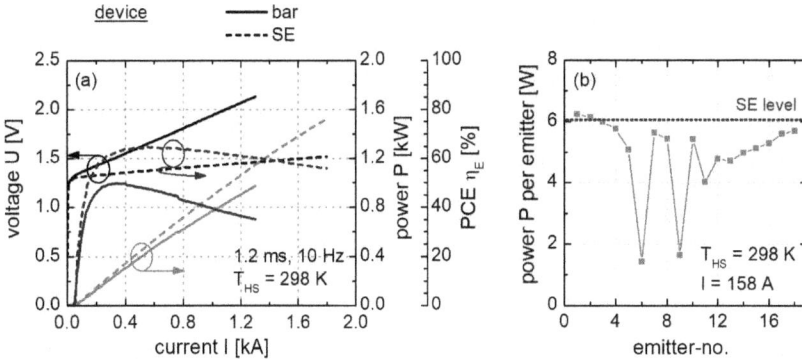

Figure 4.14: (a) Voltage U, power P and power conversion efficiency η_E versus current I at $T_{HS} = 298\,K$ for a bar (solid) and a single emitter (dashed, scaled to bar) using Design2. (b) Power distribution across the bar at a current of $I = 158\,A$.

In figure 4.14 (a) the PUI characteristics of the bar and the emitter, scaled to bar level are depicted. The threshold current of the two devices matches, but the bar suffers from a much decreased slope of the power current curve of $S = 0.91\,W/A$ compared to the single emitter's $1.00\,W/A$. Furthermore, at a current of $I = 0.76\,kA$, a kink in the bar power-current curve hints toward an emitter failure, decreasing slope further. At the maximum current of 1.3 kA, the bar emits $P = 0.97\,kW$, a significantly lower power than the expected 1.1 kW from the single emitter measurement. The possible maximum output power is given by the scaled single emitter result with 1.53 kW at $I = 1.8$ kA. A major difference between the single emitter and bar results can be seen in the voltage current characteristics. As mentioned in the analysis of the bar measurements of vertical Design1, the bar suffers from a higher series resistance due to a high package resistance. The Cu-foil establishing the n-contact acts as an in-line resistor, adding significantly to the measured voltage. Furthermore, the reduced contact area due to the patterned p-contact causes additional resistance. Unfortunately, the magnitude of this latter effect can not be differentiated from the superimposed, much bigger contribution of the package resistance. Together, the lower slope of the power curve and the massive increase in measured voltage result in a much decreased efficiency of the bar. Peak efficiency is 50% and 65% for the bar and single emitter, respectively, decreasing to 36% and 62% at $P = 0.95$ kW.

The reduced output power of the bar is confirmed on emitter level in figure 4.14 (b). Measured at a current of 0.16 kA after the high current test, the bar power is not evenly distributed across all emitters. Two partial emitter failures can be made out (emitters 6 and 9), but also other emitters show a reduced output power compared to the expected single emitter power level.

For the bar tests at lower temperatures, a second device was used with the same layout and assembly. The results of the bar tested at $T_{HS} = 223$ K under QCW conditions (1.2 ms, 10 Hz) is shown in figure 4.15. This bar had been subjected to prior high current measurements with lower pulse widths and is assumed to have been damaged in the process. Due to the complications of determining the output power at short pulses discussed in chapter 3, no reliable information on the power levels reached are available, nor are emitter failures indubitably detected, as drops in the thermoelectric power signal could be stemming from abrupt pulse shape changes. The bar measurement is compared to the scaled results from a single emitter tested at the same temperature. After the low temperature test up to the high currents, the power distribution across the bar is determined at room temperature at a low current of $I = 164$ A.

In figure 4.15 (a), the performance of the high FF bar using Design2 at $T_{HS} = 223$ K is compared to the scaled PUI-curve of a single emitter with $L = 6$ mm, $w = 200$ μm. The single emitter was measured at a comparable duty cycle (1 ms, 16 Hz) which was determined by the available current supply. Threshold current of the two devices matches but the slope of the power-current curve is much reduced for the bar which could be due to emitter damage sustained in previous measurements with shorter pulses. The lower slope together with an increased roll over compared to the single emitter lead to a maximum output power of 1.45 kW at the highest current of 1.92 kA, a 50% increase compared to the output power of 0.97 kW at the highest current reached at room temperature (1.3 kA). Contrary to the room temperature measurement, no kinks in the power current characteristics are observed during the measurement. The maximum possible output power given by the scaled SE reaches 1.72 kW at a current of 1.8 kA. Compared to the room temperature results, this increase in expected output power is mainly due to a reduced roll over of the power at higher currents and not due to an increased slope (determined at low currents). This is in line with the prior low current CW observations in which no substantial increase in external differential efficiency was observed. Major differences between single emitter and bar are again seen in the voltage current characteristic, with bar resistance much increased to $0.64\,\text{m}\Omega$ compared to $0.18\,\text{m}\Omega$ for the single emitter. This is attributed to the package resistance of the bar as well as its structured contact. The increased voltage and the lower slope lead to a peak efficiency of 53% and 70% for the bar and single emitter, respectively, with efficiency at $P = 1.45\,\text{kW}$ dropping to 30% and 63%. Even in these single emitters with long cavities, which show the lowest R_S, the efficiency at high powers is diminished significantly by the series resistance. Thus, not only does the mounting scheme of the bar have to be revised, but the series resistance of the vertical design has to be reduced without impeding the optical performance leading to the high (scaled) powers. Further design optimizations to reduce series resistance of the epitaxial layer structure will also need to analyze the reason for the increasing resistance at lower temperatures, which is not predicted from bulk layer estimations in chapter 2.

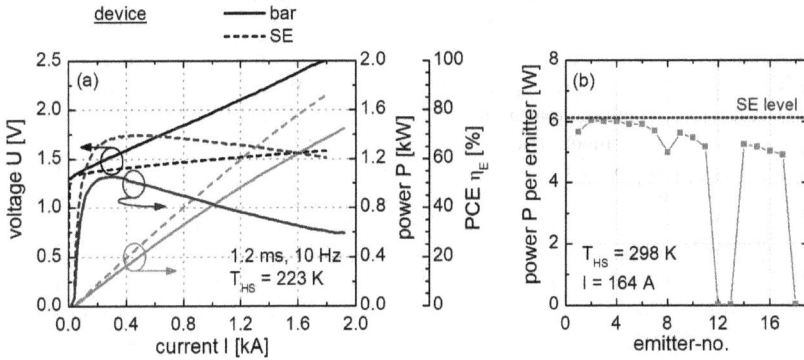

Figure 4.15: (a) Voltage U, power P and power conversion efficiency η_E versus current I at $T_{HS} = 223\,\text{K}$ for a bar (solid) and a single emitter (dashed, scaled to bar) using Design2. (b) Power distribution across the bar at a current of $I = 158\,\text{A}$ at room temperature.

After the high current, low temperature measurement, the bar was tested in a telescopic setup at room temperature to gain information on the power distribution among the emitters (figure 4.15 (b)). Measured at a current of 164 A in QCW condition (1.2 ms, 10 Hz), the power per emitter varies between 5 W and 6 W with three emitters not emitting at all. The reduced slope in figure 4.15 (b) is in accordance with the loss of three emitters detected here. Apart from the failed emitters, this bar shows a more homogenous power distribution close to the level measured from a single emitter than the bar assessed at room temperature, hinting at variations in the mounting condition between the devices. The achieved bar powers and efficiencies from the high FF bar measurements of Design2 are summarized in table 4.8 alongside further measurement values.

Table 4.8: Measurement results of high FF bars and scaled values for the SE using the vertical structure Design2.

T_{HS} [K]	$I_{max.}$ [kA]	$P_{max.}$ [kW]	$U(I_{max.})$ [V]	$\eta_E(P_{max.})$ [%]	$\eta_{E,max.}$ [%]	$\lambda(I_{max.})$ [nm]
Bar						
298	1.3	0.97	2.13	35	50	982
223	1.92	1.45	2.55	30	53	964
sc. SE						
298	1.8	1.53	1.52	56	65	982
223	1.8	1.72	1.58	61	70	958

The second bar, which was measured at the lower temperatures, is discussed in a publication [Cru13.1]. Herein, the reported power numbers had been gained by matching the slope of the QCW room temperature measurement at the established measurement station to the slope of the thermoelectric signal in the Cryo-station. As discussed earlier, this method is very prone to overestimation of the achieved power. First, the initial QCW test used a power calibration for the photodiode which inflated the slope of the bar compared to the single emitter (which serves as an upper limit for the performance of the bar) by $\sim 7\%$. Second, the change of the pulse shape over the course of the measurement was not accounted for and kinks in the detected thermoelectric signal were solely attributed to switches in the current supply. Though the reported numbers are for measurements with 1.2 ms, 10 Hz, the power values had been gained by matching the slope just above threshold to wrongly calibrated measurements with 0.1 ms, 10 Hz. By the time, the bar was measured with the longer pulse widths, it had lost emitters (as seen above) and the matching of the slope was no longer valid. At the time of the publication, these problems with the measurement setup had not been discovered and the reported values were to the best of the authors' knowledge. Retrospectively, power was overestimated by $\sim 15\%$. Nonetheless, the conclusions drawn from in the publication still hold true, as longer resonators benefit efficiency at high currents and for Design2 the power is increased at lower temperatures mainly due to reduced roll over of the power-current characteristics.

4.3 Conclusions from Iteration 0

The influence of various vertical and lateral designs on the performance of diode lasers at lower operation temperatures was explored in iteration 0. From a number of different available vertical structures, the benefits of a thinner waveguide, a higher asymmetry and a lower Al-content in the waveguide were shown experimentally. Two vertical designs were proposed for a benchmark process in which single emitters and bars with various lateral layouts were produced. The impact of stripe width and resonator length on the power and efficiency at decreasing temperatures was assessed for single emitter with Design1. While the highest powers are achieved from 4 mm long devices, longer resonators benefit from a reduced series resistance which increases efficiency at high currents. Broader stripes are shown to have lower threshold current per stripe width and higher slope efficiency, but a higher resistance times stripe width, all of which is attributed to the smaller effect of current spreading. Bars using Design1 show a lower performance than expected from single emitter measurements due to lower slope and a significant increase in voltage. The latter stems from an increased packaging resistance as well as a reduced contact area of the bar due to structuring of the wide emitters. Bars using Design2 reach very high output powers of 1.45 kW at low temperatures, but suffer from the high package resistance, emitter failures as well as a low increase in external differential efficiency. Scaled single emitter results show high output powers > 1.5 kW with efficiencies > 60% and highlight the need for low resistance bar packaging as well as low resistance vertical designs. Based on the high power single emitter measurements of Design2, the series resistance is identified as the main limit to high efficiencies at high powers. Further vertical design optimization has to be performed to reduce resistance without deteriorating the optical performance. Finally, the need for an improved measurement setup which includes tracking of the development of pulse shape and reliable power calibration for shorter pulse widths is highlighted during the test of the bars using Design2.

Chapter 5

Low Al-content Structures (Iteration 1)

In chapter 5 the design alterations based on the conclusions from chapter 4 and results of iteration 1 will be presented. The iteration focuses on increasing the efficiency of 940 nm diode lasers at the high power levels reached in chapter 4. First, options for decreasing the main limiting factor to efficiency at high powers, the electrical series resistance, are discussed and weighed on their practicability. Low series resistance vertical designs are proposed, simulated, epitaxially grown and realized in a short loop process. Diode lasers of different lengths from the short loop process are facet coated and mounted for full testing at different temperatures. The measurements provide information on the development of internal electro-optical parameters with temperature as well as the electrical properties of the designs. The results are discussed revealing amongst other factors a strong dependence of the series resistance on quantum well depth. This represents the first ever reported analysis of the reasons behind the increasing series resistance at lower temperatures.

The second part of this chapter starts by introducing the vertical and lateral designs chosen for the full laser process. The results from the short loop process are used to select the most promising variants to achieve high output powers and efficiencies at the targeted operation temperature. Measurement results are presented for single emitters and bars and compared to previous results from the short loop process. Insights are gained on the behavior at high currents which had not been accessible with the short loop processed diodes. Furthermore, low resistance packaging of bars proves to be crucial and the mounting is adjusted accordingly. The resulting assembled bars sustain for the first time efficiencies $> 55\%$ up to power levels of 2 kW. In an effort to reduce excess resistance, changes to the active zone are made and bar performance is shown to be enhanced. Finally, the dissipated heat of single emitters at low temperatures and high powers is broken down into the contributing loss mechanisms and the most important loss channel is identified for further design optimization.

5.1 Design and Optimization of Low Resistance Structures

After the initial design benchmarking of iteration 0 the focus is shifted to a more detailed experimental assessment of the design changes needed to increase the efficiency of the diode lasers while sustaining the high power levels especially seen in the scaled single emitter results. As discussed at the end of chapter 4 the series resistance R_S is the main limit to high efficiencies at high powers in room temperature optimized designs. Thus, the focus is put on reducing R_S without impairing the other characteristic parameters. The vertical structure Design1 from chapter 4 was chosen as a starting point. This structure uses a 2.4 µm thick waveguide that provides a low fast axis beam divergence (vertical far field distribution with 95% power content $\Theta_{95\%}^v = 35°$) potentially beneficial for pumping application [Ert11]. Furthermore, the thicker waveguide reduces the facet load, promising higher catastrophic optical mirror damage (COMD) thresholds [Bot99].

5.1.1 Options to reduce series resistance

The vertical structure of a diode laser is simplified as a stack of individual bulk layers in figure 5.1.

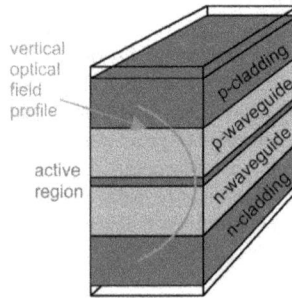

Figure 5.1: Vertical design of a diode laser, simplified as a stack of individual layers.

The total series resistance R_S of the stack can be estimated by adding up the contributions of each layer j. If current spreading and interface effects can be neglected, the contribution of a single layer can be calculated with the knowledge of its thickness d_j, the contact area (width w and length L), the free electron and hole densities and mobilities n_j, p_j, $\mu_{e,j}$, and $\mu_{h,j}$ respectively:

$$R_j = d_j / [wL(n_j e \mu_{e,j} + p_j e \mu_{h,j})]. \tag{5.1}$$

For large area devices such as broad area lasers, current spreading effects play only a minor role [Cru13] and equation 5.1 typically matches low current measurements at room temperature. Generally the biggest contribution to the total series resistance is made up by the waveguide layers. Here, n_j and p_j are small as the doping is kept low to avoid losses by free carrier absorption calculated for each layer j by

$$\alpha_{i,j} = \Gamma_j [\sigma_h p_j + \sigma_e n_j] \tag{5.2}$$

where Γ_j is the overlap of the optical mode with the layer j and σ_h and σ_e the absorption cross sections for holes and electrons, respectively. For the vertical structure Design1 from iteration 0 for example, the total resistance of the layers for a $w = 100\,\mu m$ wide and $L = 4\,mm$ long stripe adds up to $R_S = \sum_j R_j = 15\,m\Omega$ using values for $\mu_{e,j}$ and $\mu_{h,j}$ from [Sot00] with an additional $5\,m\Omega$ included to account for the $120\,\mu m$ thick substrate itself. In table 5.1, the layers are summarized by their function and the values for R_S and α_j for the heterostructure are listed. The optical loss in the quantum well is caused by an estimated carrier density of $N = 1 \times 10^{18}\,cm^{-3}$.

Table 5.1: Contributions of layers to the series resistance and optical loss for a $100\,\mu m$ wide, $4\,mm$ long stripe for a vertical design used in iteration 0.

Design1		
Layer	Resistance [mΩ]	Optical loss [cm^{-1}]
p-cladding	0.7	0.01
p-waveguide	12.7	0.22
active zone	0	0.06
n-waveguide	1	0.18
n-cladding	0.1	0.09
substrate	5	0
	19.5	0.5

As predicted, the waveguide layers contribute the most to the total of $R_S = 19.5\,m\Omega$, accruing 70% of the design's resistance. Especially the large contribution of the p-waveguide (65% of total R_S) poses a challenge for low resistance structures. The optical absorption cross section of holes σ_h is reported in [Pet07] to be $12\,cm^{-1}$, three times as big as the one for electrons σ_e and dictates a very low p-doping in the optical confinement layers. Overall, the electrical resistance of the waveguide is the main limit to low R_S structures.

Looking at equation 5.1 the options for reducing R_S especially in the waveguide layers as well as the side effects are as follows:

1. increase contact area wL. However, longer laser cavities increase the amount of wafer material used, making the diode lasers more expensive. Using wider stripes has a limited benefit for the targeted high fill factor bars, in which most of the bar's width is covered by emitters.

2. reduce the layer thickness d_j. However, implementing thinner waveguides generally leads to a higher facet load d/Γ and thus increases the risk of COMD [Bot99]. Furthermore, an increased vertical far field angle $\Theta_{95\%}^v$ can be limiting for pumping applications [Ert11].

3. increase the free carrier densities n_j and p_j. However, a higher doping in the waveguide layers would increase the optical absorption and thus increase the in-

ternal optical loss, limiting optical output power and increasing threshold current, potentially eliminating efficiency increases.

4. increase carrier mobilities $\mu_{e,j}$ and $\mu_{h,j}$. Looking at [Sot00], a lower aluminum content in an $Al_xGa_{1-x}As$ bulk layer increases the mobility. However, the use of low Al-content waveguides reduces the barrier height around the quantum well leading to an increase of thermal leakage of carriers into the waveguide layers.

The option of reducing the Al-content in the waveguide layers proves to be the most promising to achieve higher efficiencies as the deteriorating effect of the lower barriers around the active zone is mitigated by a lower heat sink temperature T_{HS}. This effect is illustrated in figure 5.2, where the conduction band edge in and around the active zone is plotted as well as the confined electron state gained from solving the Schrödinger equation. The absolute barrier height ΔE is defined as the energy difference between the waveguide conduction band edge and the confined electron state.

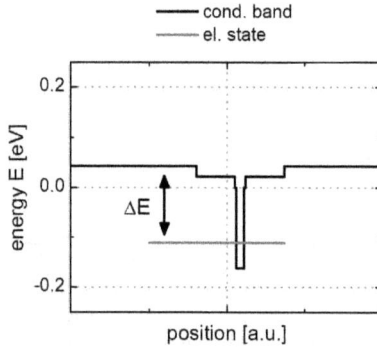

Figure 5.2: Conduction band edge (black) and confined electron stated (red) in the vicinity of the quantum well.

While ΔE does not significantly change with temperature, the thermal energy $E_{th} = k_B T_{HS}$ of the bound carriers depends strongly on the temperature. Together, this leads to an increase in *effective* barrier height $\Delta E/k_B T_{HS}$. Thus, lower barriers are acceptable for low temperature operation whereas room temperature operation will be degraded. Consequently designs had to be found with an optimum Al-content in the waveguides to provide as low of a series resistance as possible without deteriorating the performance at a heatsink temperature of $T_{HS} = 200\,K$.

5.1.2 Vertical design of low resistance structures

Studies on the impact of low Al-content waveguides on the electrical and power characteristics are performed using the baseline structure Design1 from iteration 0. The original structure had been optimized for high power room temperature operation and featured an $Al_{0.15}Ga_{0.85}As$ waveguide. Consecutively the Al-content in the waveguide is changed between $x = 17.5\%$ and 0% to probe limits. Furthermore, following [Cru13] the Al-content in the n-cladding is to remain 4% higher than in the waveguide to suppress higher order modes and has to be changed accordingly. The composition of the other layers is not altered. The thickness of all layers and their doping profiles remains fixed throughout as does the composition and position of the InGaAs quantum well.

The resulting vertical structures were analyzed using the vertical mode solver QIP [Wen90]. In figure 5.3 (a) the band structure is shown for the three example configurations with $x = 0\%$, 7.5% and baseline 15% Al-content in the waveguide at $T_{HS} = 298\,K$. Here, the changes and their effects on the effective barrier height are clearly visible: while the band structures are identical in the p-cladding, the band gap shrinks for lower Al-contents in the waveguide and the n-cladding. As the bound energy state in the quantum well stays roughly constant between the structures, subsequently, the quantum well depth changes and the barrier height ΔE gets smaller. The effective barrier height $\Delta E/k_B T_{HS}$ decreases accordingly.

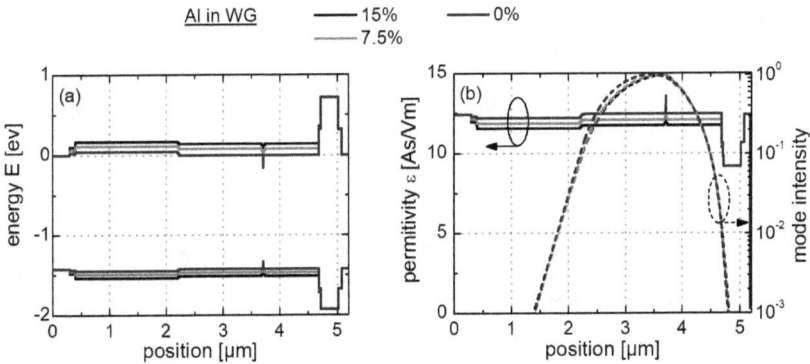

Figure 5.3: Calculations for designs with $x = 15\%$ (black), 7.5% (red) and 0% (blue) in the waveguide. (a) Band structure. (b) Permitivity (solid) and fundamental mode (dashed).

The permitivity as well as the normalized intensity of the fundamental mode are shown in figure 5.3 (b) for the three Al-contents. The lower the Al-content in the waveguide, the stronger the mode overlaps with the n-waveguide decreasing the fraction of modal intensity in the active region. Lower Al-content structures thus have a lower confinement factor Γ. The calculated values for Γ, the series resistance R_S for a 4 mm long and 100 µm wide stripe, the internal optical loss α_i and the barrier height ΔE are summarized in table 5.2.

Table 5.2: Calculated values for vertical structures with different Al-content in the waveguide. R_S is given for a 100 µm wide, 4 mm long single emitter and includes 5 mΩ of substrate and contact resistance. The design with 15% Al-content serves as a reference.

Al [%]	Γ [%]	R_S [mΩ]	α_i [cm^{-1}]	ΔE [meV]	$\Delta E/k_B T_{HS}(300\,\text{K})$	$\Delta E/k_B T_{HS}(200\,\text{K})$
0	0.32	16.3	0.47	91	3.5	5.3
4	0.354	17	0.48	127	4.9	7.4
7.5	0.369	17.7	0.48	171	6.6	9.9
10	0.371	18.2	0.48	193	7.5	11.2
12.5	0.375	18.8	0.49	213	8.2	12.4
15	**0.378**	**19.5**	**0.49**	**232**	**9**	**13.5**
17.5	−	20.2	0.49	252	9.7	14.6

The barrier height of $\Delta E = 232$ meV of the room temperature optimized reference structure corresponds to an effective barrier height of $\Delta E/k_B T_{HS} = 9$ at $T_{HS} = 298$ K. At $T_{HS} = 200$ K the 7.5% Al-content design has an effective barrier height of $\Delta E/k_B T_{HS} = 9.9$, suggesting that a high power operation of this structure is feasible without degradation to the optical performance. Even the more aggressive 4% Al-content structure has the potential to perform well at the low temperatures, while the barrier height of the pure GaAs waveguide design is expected to suffer heavily from thermal leakage even at the lowest temperatures due to its low barriers.

For the two structures with the lowest Al-contents in the waveguide ($x = 0\%$ and $x = 4\%$) a design change to the active zone was also explored in which the single quantum well was replaced by a double quantum well. The properties of all other vertical layers (doping, thickness, Al-content) are kept constant. The conduction band and the bound electron state in the vicinity of the active region are depicted in figure 5.4. The additional quantum well results in a lower carrier concentration in each quantum well, thus reducing leakage into the waveguide [Pie09] at the cost of a higher threshold current. Consequently, a lower peak efficiency can be expected but the structure should have an increased differential internal efficiency η_i and sustain a high efficiency up to high currents due to the lower roll over.

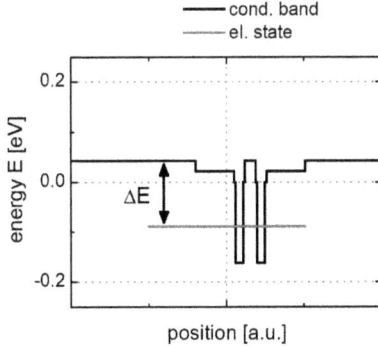

Figure 5.4: Band structure for the $x = 0\%$ design with a double quantum well.

5.1.3 Short pulse measurements of designs with low Al-content in the waveguide

The proposed designs listed in table 5.2 as well as the two designs with the double quantum wells are grown on 3 inch GaAs wafers using MOVPE and subsequently processed in a short-loop process (SLP). Short pulse measurements ($1.5\,\mu s$, $5\,kHz$) of uncoated, unmounted bars with varying resonator lengths are performed at $T_{HS} = 293\,K$. Following standard practice [Erb00], the internal parameters can be extracted from the dependence of threshold current and slope on the resonator length, as explained in chapter 2. In figure 5.5 (a) the inverse of the external differential efficiency η_d is plotted against the chip length for all the SQW designs with $100\,\mu m$ stripe width while figure 5.5 (b) displays the threshold current density versus length. The design with $x = 0\%$ aluminum in the waveguide does not lase at this temperature up to the maximum current of $I = 2\,A$. From figure 5.5 (a) and using equation 2.15 it is clearly visible how the differential internal efficiency η_i deteriorates with decreasing Al-content, as the intersection of the linear fit to the points with the ordinate increases. This is attributed to the decrease of the quantum well depth and a decrease in ΔE which leads to poor performance of low Al-content structures at room temperature. At the same time the slope of the fit α_i/η_i changes accordingly with the decreasing η_i in figure 5.5 (a) meaning that internal loss α_i is constant across the structures as expected from the universal doping profile. The only design showing a visibly higher slope is the $x = 4\%$ structure. This can be explained by looking at figure 5.5 (b), in which a similar trend to the behavior of η_i can be identified, as the transparency current density J_T calculated from equation 2.9 increases for lower Al-contents. The biggest increase in J_T can be expected between $x = 7.5\%$ and 4% in the waveguide. Therefore, the number of carriers in the quantum well is much higher in the lower Al-content case, leading to higher absorption in the active region according to [Kau16]. The modal gain Γg_0 calculated from equation 2.9 only changes slightly which is due to the lower confinement factor of low Al-content structures, as the fundamental mode shifts more towards the n-waveguide as seen in simulations and in figure 5.3 (b).

71

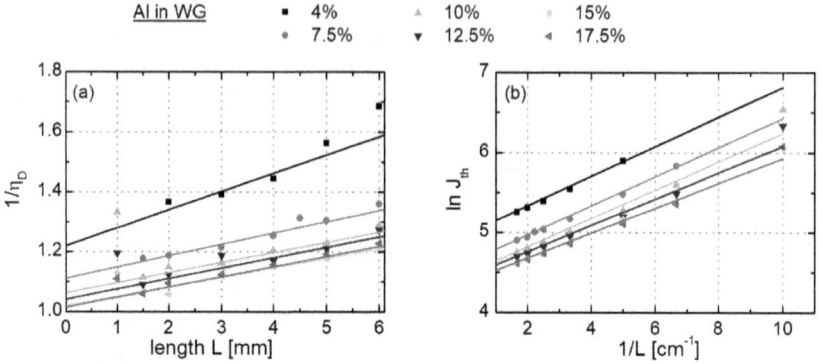

Figure 5.5: Results from short pulse measurements performed at $T_{HS} = 293\,\mathrm{K}$ for structures with different Al-content in the waveguide. Length dependence of (a) external differential efficiency and (b) threshold current.

The observations made from the graphical analysis of figure 5.5 are put into numbers in table 5.3.

Table 5.3: Internal parameters for vertical structures with different Al-content in the waveguide with 100 µm stripe width, measured at $T_{HS} = 293\,\mathrm{K}$. T_0 values are determined for a 2 mm long device.

Al [%]	η_i [%]	α_i [cm^{-1}]	J_T [A/cm^2]	Γg_0 [cm^{-1}]	T_0 [K]
0					
4	82	0.6	130.2	6.5	33
7.5	90	0.41	93.8	6.6	58
10	94	0.38	82.4	6.8	78
12.5	96	0.4	79.4	7.3	90
15	98	0.37	75	7.7	110
17.5	98.5	0.4	74.9	7.7	124
0 (DQW)	73	0.61	307	9.8	47
4 (DQW)	95	0.85	190.4	14.2	87

The extracted values show the strong dependence of η_i and J_T on the Al-content in the waveguide. However, these measurements were conducted at room temperature, so the low Al-content structures were expected to show poor performance. Unfortunately, the measurement setup did not allow for short pulse measurements of unmounted diode bars to be carried out at temperatures below $T_{HS} = 288\,\mathrm{K}$. Furthermore, differences in the electrical characteristics of the different structures are not visible in the measurement as the voltage is only determined at $I = 0.1\,\mathrm{A}$ and $2\,\mathrm{A}$ with a $10\,\mathrm{mV}$ resolution. Thus, no experimental information could be gained on the influence of the Al-content on the series resistance. Low temperature measurements as well as reliable voltage measurements were therefore only possible with fully assembled diode lasers.

It can already be seen from the T_0 values in table 5.3 that the structures featuring a double quantum well possess a better temperature sensitivity than the single quantum well designs with the respective Al-content in the waveguide, meaning that lasing is observed even for the lowest Al-content DQW design. Furthermore, the values of η_i are still high at room temperature compared to the respective single quantum well designs. For each design $4\,\mathrm{mm}$ bars from the SLP were cleaved and facet coated with front and back reflectivities of $R_f = 2\%$ and $R_r = 95\%$, respectively. Additionally, $3\,\mathrm{mm}$ and $6\,\mathrm{mm}$ bars with the $x = 12.5\%$ design were coated the same way to perform length dependent measurements at lower temperatures. Facet passivation [Res05] was omitted in all cases, as the diodes were not to be tested for durability and the lower COMD threshold limited output powers to $\sim 10\,\mathrm{W}$. Subsequently, three $100\,\mu\mathrm{m}$ wide devices per design and length were hard soldered p-side up onto CuW-submounts using AuSn. Due to the aluminum used for the p-side metallization in the SLP, p-side down mounting was not possible and contacting the p-side was done using aluminum bond wires.

5.1.4 Low Temperature Length-Dependent Measurements

For the designs with $x = 12.5\%$ in the waveguide, $100\,\mu\mathrm{m}$ emitters with resonator lengths of $L = 3, 4$ and $6\,\mathrm{mm}$ had been assembled. Three lasers per length and design were tested at room temperature under quasi-continuous wave (QCW) condition ($1.2\,\mathrm{ms}$, $10\,\mathrm{Hz}$) and the threshold current and external slope efficiency are extracted by fitting the power-current characteristics in the interval $[I_{th} + 0.5\,\mathrm{A}, I_{th} + 2.5\,\mathrm{A}]$. The standard error of the mean in threshold current and external differential efficiency is $< 0.5\%$.

The comparability between the short pulse and the QCW measurement is explored in figure 5.6. The values from the length dependent QCW measurements of exemplary assembled devices are compared to the short pulse measurements presented in figure 5.5. The inverse of the external differential efficiency η_d and threshold current density J_{th} are plotted against the mirror loss to establish comparability of coated and uncoated devices. The slightly different temperatures of $T_{HS} = 293\,\mathrm{K}$ for the short pulse and $288\,\mathrm{K}$ for the QCW measurement are not expected to cause significant differences in performance.

In figure 5.6 (a) the dependence of external differential slope efficiency on the mirror loss is shown for the unmounted and fully assembled devices. The fits to the measurement points run parallel but are shifted on the vertical axis. For the internal parameters derived from the fit this means that internal loss is the same ($0.4\ \mathrm{cm}^{-1}$ for the unmounted, $0.42\ \mathrm{cm}^{-1}$ for the mounted diodes) but the internal differential efficiency is different (96% for the unmounted, 86% for the mounted diodes). The difference in η_i can be explained with the power calibration of the measurement station for the uncoated, unmounted devices. Its focus is the qualitative assessment of vertical structures and thus does not produce exact power values which translates into an overestimation of slope and thus results in an exaggerated internal differential efficiency. Compared to the room temperature pulsed measurement setup, a careful power calibration is performed at the low temperature measurement station as detailed in chapter 3 and power values are referenced to national standards. Thus, the internal differential efficiency numbers from table 5.3 are adjusted using the relative difference between the measurement stations and are given in table 5.4.

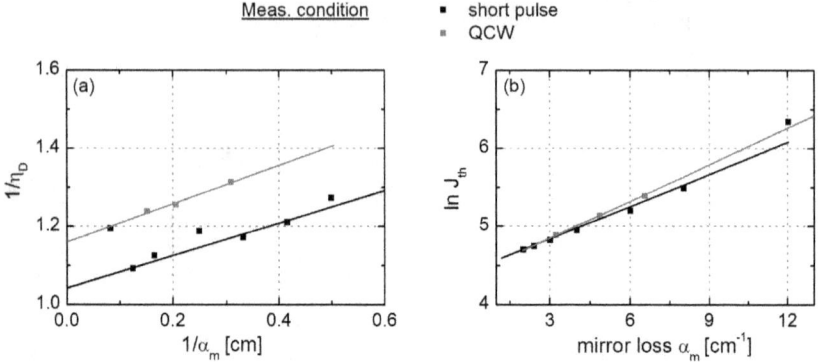

Figure 5.6: Results from length dependent measurements of mounted (QCW) and unmounted (short pulse) devices for the design with 12.5% Al-content in the waveguide. (a) External differential efficiency and (b) threshold current density are plotted against the mirror loss.

Table 5.4: Internal efficiency from table 5.3 and adjusted values based on measurements of mounted diodes conducted with accurate power calibration. In bold are listed the Al-content structure used to calculate the adjustment factor.

Al [%]	$\eta_{i,uncal.}$ [%]	$\alpha_{i,adj.}$ [%]
0		
4	82	73
7.5	90	81
10	94	84
12.5	**96**	**86**
15	98	88
17.5	98.5	88
0 (DQW)	73	65
4 (DQW)	95	85

The analysis of mirror loss dependence of the threshold current is shown in figure 5.6 (b). Good agreement can be seen in the vertical position of the two fits, corresponding to a similar transparency current density ($79.4\,\mathrm{A\,cm^{-1}}$ for the unmounted, $74.7\,\mathrm{A\,cm^{-1}}$ for the mounted diodes). The modal gain determined from the mounted and unmounted diodes differs slightly ($7.3\,\mathrm{cm^{-1}}$ for the unmounted, $6.4\,\mathrm{cm^{-1}}$ for the mounted diodes), as can be seen from the different slope of the fit. Generally, the two measurements are comparable (with the difference in η_i in mind) and conclusions drawn from the unmounted devices are thus expected to be reproduced in the coated, fully mounted diodes.

After the comparability between the two measurement methods is established, one exemplary device for the $x = 12.5\%$ Al-content design per length is tested at temperatures between $T_{HS} = 218 \dots 298\,\mathrm{K}$ in $10\,\mathrm{K}$ steps using the first low temperature measurement setup presented in chapter 3. The devices are operated under QCW condition ($1.2\,\mathrm{ms}$, $10\,\mathrm{Hz}$) and the heat sink temperature is controlled at the CuW-submount. The values for $\eta_d(T_{HS})$ and $I_{th}(T_{HS})$ are gained by fitting the power-current characteristics in the interval $[I_{th} + 0.5\,\mathrm{A}, I_{th} + 2.5\,\mathrm{A}]$ and applying the standard error of the mean gained earlier from the room temperature measurement. The values for the three lengths and the nine temperatures for $x = 12.5\%$ design are plotted in figure 5.7 and the extracted numerical values for the internal parameters are plotted against T_{HS} in figure 5.8.

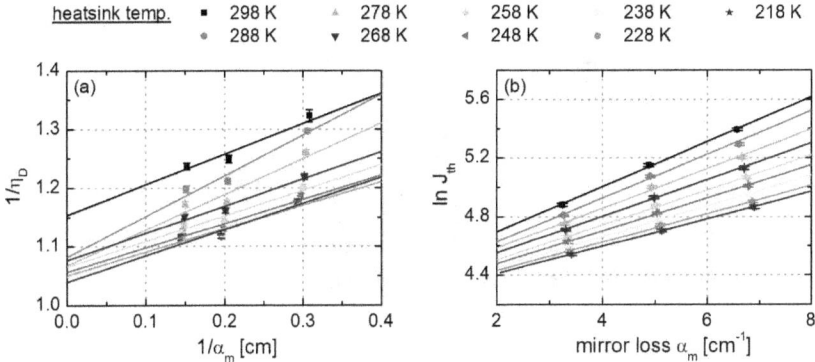

Figure 5.7: Measurement values for the devices with the $x = 12.5\%$ design for different lengths and T_{HS}. (a) External differential efficiency and (b) threshold current are plotted against the mirror loss.

From figure 5.7 (a) the development of the differential internal efficiency with temperature is clearly visible. As T_{HS} drops, η_i increases (intersection of the fit with ordinate decreases) while no clear trend is observed for the slope of the fit (leading to the internal optical loss). The modal gain Γg_0 increases with lower T_{HS} while J_T decreases as can be seen from the slope and ordinate intersection of the fit in figure 5.7 (b), respectively.

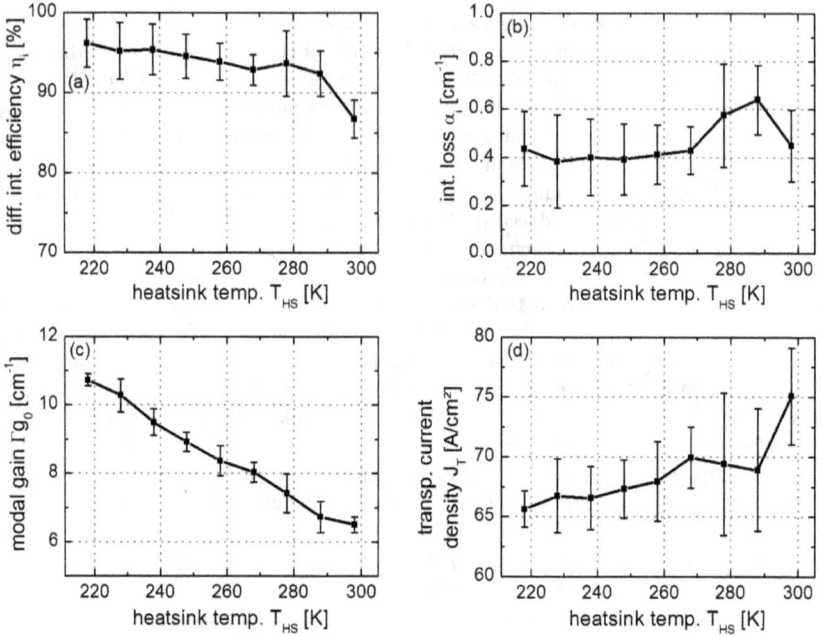

Figure 5.8: Development of internal parameters of the $x = 12.5\%$ design with T_{HS}. (a) Differential internal efficiency η_i, (b) internal optical loss α_i, (c) modal gain Γg_0 and (d) transparency current density J_T.

The error bars in figure 5.8 originating from the error of the linear fit over the three lengths. In numbers, η_i shows a big improvement from 86% to 96% (figure 5.8 (a)) when going from $T_{HS} = 298$ to 218 K. The strong rise in the temperature interval from 298 to 278 K is attributed to the reduced leakage of carriers out of the active zone as well as to the freeze out of non-radiative recombination channels. No clear temperature dependence of the internal optical loss can be extracted from figure 5.8 (b), but instead α_i stays roughly constant around $0.4\,\mathrm{cm}^{-1}$ over the entire temperature range. This observation is in contrast to the temperature dependence of α_i suggested in [Wen10], although errors are fairly large. The rise of Γg_0 from $6.5\,\mathrm{cm}^{-1}$ at 298 K to $10.7\,\mathrm{cm}^{-1}$ at 218 K (figure 5.8 (c)) is attributed to the narrower gain spectrum at the lower temperature, as discussed in chapter 2. This leads to an increase in material gain g_0 at the lasing wavelength, and while the confinement factor Γ is approximately independent of temperature, this translates into an increase in Γg_0 [Wen10]. The suppressed leakage and the lower non-radiative recombination not only benefit η_i but J_T also decreases from $75\,\mathrm{A\,cm}^{-1}$ at $T_{HS} = 298$ K to $66\,\mathrm{A\,cm}^{-1}$ at 218 K (figure 5.8 (d)). Generally, a clear improvement across the internal parameters can be seen when lowering the heatsink temperatures, with only α_i staying constant.

5.1.5 Measurement of Al-content Matrix

After establishing the benefit of low-temperature operation to the internal parameters, the dependence of the performance on the Al-content in the waveguide was tracked for different temperatures. The assembled $L = 4$ mm and $w = 100$ μm single emitters of the remaining designs ($x = 0\%$, 4%, 10%, 15%, 17.5%) were tested in the same temperature range of $T_{HS} = 298$ to 218 K under CW condition up to a current of $I = 5$ A. In the first low temperature measurement setup, the best power calibration was available for continuous wave operation as stated in chapter 3. At the lowest temperature, some higher currents could not be reached as the cooling setup was not able to remove the waste heat. Also, measurements were aborted when roll over became too strong (as in the case for the $x = 4\%$ structure at $T_{HS} = 288$ K for example). To separate the change of the energy gap E_g with temperature from the structure dependent voltage, the defect voltage U_d is calculated, defined as the voltage in excess of the voltage needed to overcome the band gap. Above threshold, it can be expressed as a linear function in I, similar to equation 2.17.

$$U_d = U - hc/q\lambda = U_{d,0} + R_S \cdot I \qquad (5.3)$$

with h the Planck constant, c the speed of light and $U_{d,0}$ as the ordinate intersection. Exemplary curves for the power-current characteristic as well as the defect voltage U_d are shown in figure 5.9 at the temperatures of 288, 258, 238 and 218 K for Al-contents of $x = 0\%$, 4%, 10% and 15%. At $T_{HS} = 288$ K, the design with $x = 15\%$ reaches the highest output powers (figure 5.9 (a)), while the lasers with $x = 10\%$ in the waveguide shows a slightly higher threshold current and a degraded slope. The low Al-content designs perform poorly as seen by the high threshold current, low slope and strong roll over experienced by the $x = 4\%$ structure. The design with pure GaAs in the waveguide does not even lase at this temperature. At the same time as a lower Al-content degrades room temperature optical performance, it leads to a benefit in the electrical behavior as the defect voltage is significantly reduced for low Al-content structures. Operating the lasers at $T_{HS} = 258$ K, the difference in the power-current characteristics between the $x = 15\%$ and $x = 10\%$ design becomes less pronounced (figure 5.9 (b)), as threshold current and slope are very similar (the lower Al-content design only shows a deviation in power at higher currents due to roll over). Higher absolute powers are reached by all structures as I_{th} is decreased and S is higher. The $x = 4\%$ design is now able to sustain currents up to 5 A but has a visibly increased threshold current and lower slope compared to the high Al-content designs. At this temperature, the pure GaAs structure reaches lasing, however at high currents of ~ 2 A and with poor slope. The defect voltage at 5 A increases compared to the room temperature measurement for all structures, hinting at either an increase in $U_{d,0}$ or R_S at this lower temperature. The positive impact of low Al-contents on U_d is again visible at this temperature. However, there is only little improvement seen in U_d when going from $x = 4\%$ to 0%. Lowering the temperature further to $T_{HS} = 238$ K, the optical performance between the two higher Al-content structures becomes even more similar with threshold current decreasing and slope increasing further (figure 5.9 (c)), eventually reaching powers > 5 W at 5 A. The performance gap to the $x = 4\%$ design gets smaller and the $x = 0\%$ design now reaches output powers of > 3 W still suffering from a significantly higher threshold current and lower slope. Again, the defect voltage increases in all cases and little difference is seen in the defect voltage-current characteristics of the two low Al-content designs. This is the only temperature, at which all structures reach the maximum current of 5 A which permits a comparison of U_d at this point. The $x = 4\%$ design has a defect voltage of 93 mV, almost half of the 174 mV for the 15% design and still significantly less than the 126 mV for the 10% structure. The design with GaAs in the waveguide exhibits $U_d = 80$ mV at this current. As seen by the slope of the U_d curve, the series resistance is the main reason for the decrease of the defect voltage at 5 A.

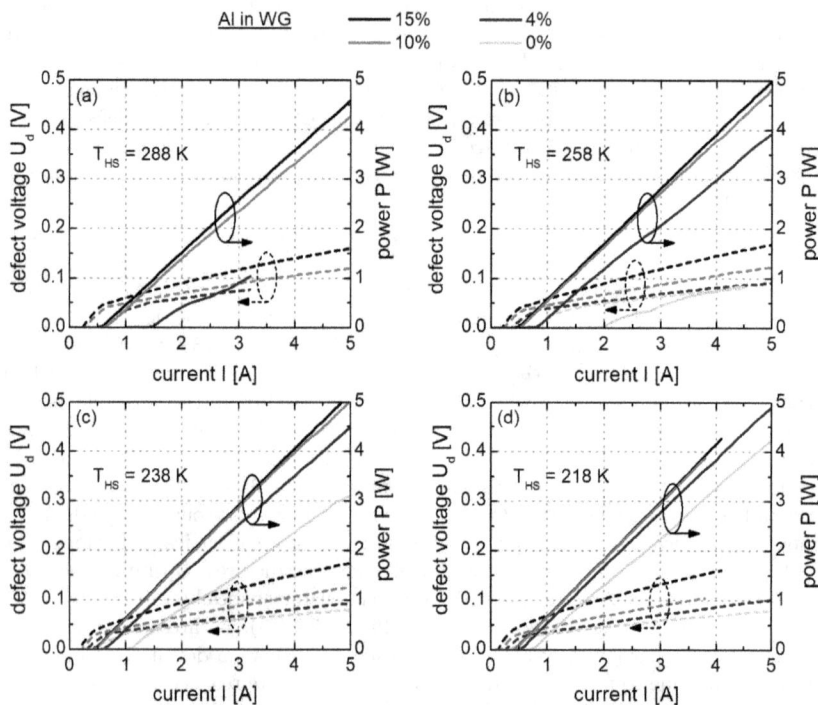

Figure 5.9: Power (solid) and defect voltage (dashed) versus current of designs with different Al-content in the waveguide at (a) T_{HS} = 288 K; (b) T_{HS} = 258 K; (c) T_{HS} = 238 K; (d) T_{HS} = 218 K.

Finally, at T_{HS} = 218 K no difference in threshold current and slope of the $x = 15\%$ and 10% design can be observed (figure 5.9 (d)). The $x = 4\%$ structure only shows marginally higher threshold current and a slightly degraded slope, with the lowest Al-content design still experiencing significant deterioration. Electrically, the pure GaAs structure shows the best performance as the defect voltage is the lowest of all four structures. However, the $x = 4\%$ design only has a slightly higher U_d which does not outweigh the substantially better optical performance.

As for all assembled single emitter devices, the characteristic parameters I_{th}, η_d and R_S are extracted by linearly fitting the power-current-voltage characteristics in the interval $[I_{th} + 0.5\,A, I_{th} + 2.5\,A]$. This is done for all structures and all measured temperatures. The results are shown in figure 5.10 for selected temperatures of T_{HS} = 288 K, 258 K and 218 K.

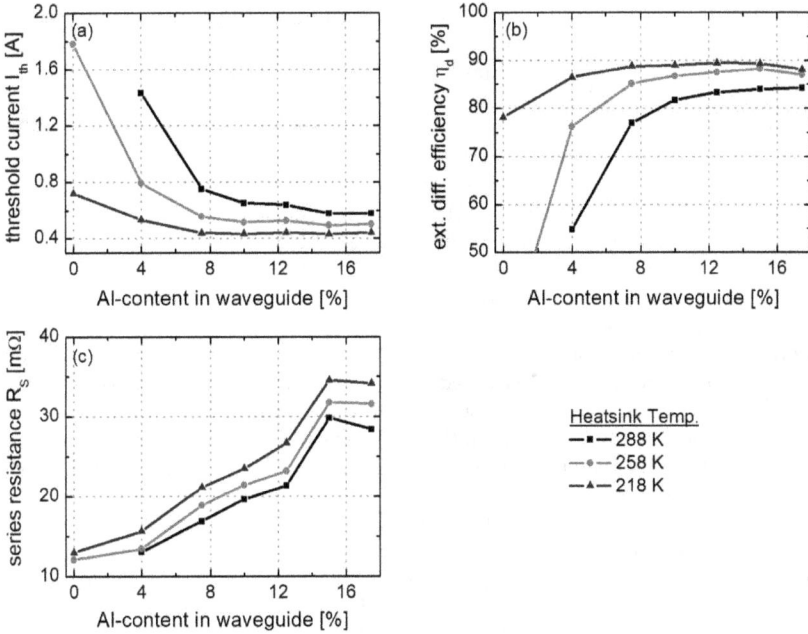

Figure 5.10: Extracted characteristic parameters versus Al-content in the waveguide at 288 K (squares), 258 K (circle) and 218 K (triangle). (a) Threshold current I_{th}; (b) external differential efficiency η_d; (c) series resistance R_S.

In figure 5.10 (a), the threshold current is plotted against the Al-content in the waveguide. For a temperature of T_{HS} = 288 K, a strong dependence of I_{th} on the Al-content is visible. As explained previously for the figure 5.9, a lower Al-content significantly increases the threshold current at room temperature. This effect becomes less pronounced when lowering the temperature and at T_{HS} = 218 K, all structures with an Al-content of $x \geq 7.5\%$ have a universal threshold current of $\sim 440\,mA$. For $x = 4\%$ and 0%, threshold current is still higher with 530 mA and 720 mA, respectively. However, it is expected that the effect of increasing independence of I_{th} of Al-content with lower temperature continues and that for the targeted temperature of T_{HS} = 200 K the $x = 4\%$ structure inhibits the same threshold current as higher Al-content structures. The dependence of Al-content on the external differential efficiency shows a similar behavior (figure 5.10 (b)). While η_d depends significantly on the Al-content in the waveguide at T_{HS} = 288 K,

79

it becomes less evident at lower temperatures. At T_{HS} = 218 K, a universal external differential efficiency of $\sim 89\%$ is observed for all structures with $x \geq 7.5\%$. Again, η_d is degraded for designs with lower Al-content for the lowest measured temperature, but for T_{HS} = 200 K η_d is predicted to take a value of $\sim 90\%$ down to $x = 4\%$. Finally, the benefit of lower Al-contents in the waveguide on the series resistance is shown in figure 5.10 (c). R_S drops from 30 mΩ for the original $x = 15\%$ structure down to 13 mΩ for $x = 4\%$ at T_{HS} = 288 K. Keeping in mind the fixed 5 mΩ contribution of the substrate, this corresponds to a reduction of R_S by 68%. The strong reduction in R_S for lower x is true for every measured temperature. There is little benefit to R_S in further lowering the Al-content to $x = 0\%$, as seen for the values at T_{HS} = 258 K and 218 K. For every Al-content, however, the series resistance increases with decreasing temperature. This behavior is not in accordance with theory, which predicts higher mobilities at lower temperatures and thus lower resistance (chapter 2). Furthermore, it can be seen that the increase in R_S from T_{HS} = 288 K to 218 K is more pronounced at higher Al-contents. For example, comparing the values at 288 K and 218 K, $\Delta R_S = 5$ mΩ for $x = 15\%$ and $\Delta R_S = 2.5$ mΩ for $x = 4\%$.

After the analysis of the characteristic parameters, the $x = 4\%$ design promises the best performance at the targeted temperature of T_{HS} = 200 K. No deterioration of optical performance is expected compared to the room-temperature optimized designs while the electrical performance is clearly enhanced due to the reduced series resistance. However, an unexplained increase in series resistance is observed when lowering the temperature whose magnitude is also dependent on the Al-content of the design. This suggests that other factors than the pure mobility increase play a role at lower temperatures and will be further investigated.

Figure 5.10 showing the measured characteristic values in dependence of the Al-content in the waveguide offers a phenomenologic view of the devices' behavior at different temperatures. To analyze the underlying effects of the decreasing threshold current with temperature and Al-content, for example, a different representation is necessary. In section 5.1.3 it was suggested that I_{th} and η_d are affected by thermionic leakage out of the quantum well and are thus dependent on the barrier height ΔE. Also, the characteristic parameters still inherit device dependent contributions which hinder a comparison between different structures. For example, the threshold current depends partly on the modal gain Γg_0 as shown in chapter 2 and visible from equation 2.8, and Γg_0 is different for all the structures due to a varying confinement factor Γ as seen in table 5.3. The same holds true for the external differential efficiency (equation 2.14), which has contributions from α_m which depends on the wavelength and thus on the temperature. Thus, I_{th} and η_d will be replaced by J_T and η_i, respectively, and their representation in dependence of the barrier height ΔE is sought.

First, the relative barrier height $\Delta E/k_B T_{HS}$ is calculated for each value of x and T_{HS}. Second, $\Gamma g_0 (T_{HS}, x)$ is determined. Since all devices tested use the same active region in which the well composition and thickness are fixed, the dependence of Γg_0 on T_{HS} is assumed to be structure independent, although barrier design generally *can* impact the band structure in the well and thus affect g_0. The measured behavior of Γg_0 in figure 5.8 (c) for the $x = 12.5\%$ structure can thus be transferred to the other designs. The difference in Γg_0 between the structures had already been measured and is listed in table 5.3. Combining the measured *structural* Γg_0 at $T_{HS} = 293\,\text{K}$ with the measured *temperature dependent* Γg_0 for the $x = 12.5\%$ structure leads to $\Gamma g_0\,(T_{HS}, x)$. The internal loss α_i is assumed to be dominated by free-carrier absorption outside the active region in accordance with [Pet07]. Since all structures use the identical vertical doping profile, the conclusions of figure 5.8 (b) can be used and α_i is fixed at $0.4\,\text{cm}^{-1}$ for all structures and temperatures. Finally, the temperature dependence of α_m is determined by measuring the wavelength-dependent front facet reflectivity R_f of coating samples at room temperature (R_r is taken as constant). As the refractive index of the semiconductor and the dielectric layers only changes minimally with temperature, the mirror loss at each temperature can be calculated from measured lasing wavelength and the determined $R_f\,(\lambda)$ dependence. Altogether, the structure dependent values for J_T and η_i can now be gained from equations 2.8 and 2.14, respectively, for each structure and temperature. They are separated from universal effects (in Γg_0 and α_m) and can be displayed solely dependent on $\Delta\,E/k_B T_{HS}$. The values for the transparency current density and the differential internal efficiency as well as the series resistance are plotted against $\Delta E/k_B T_{HS}$ in figure 5.11 for the different structures. The points correspond to the previous measurement range of $T_{HS} = 298\,\text{K}$ to $218\,\text{K}$ for all structures except for the $x = 4\%$ ($T_{HS} = 288\,\text{K}$ to $218\,\text{K}$) and $x = 0\%$ devices ($T_{HS} = 258\,\text{K}$ to $218\,\text{K}$).

In figure 5.11 (a), the calculated differential internal efficiency η_i for all structures is plotted against the relative barrier height $\Delta\,E/k_B T_{HS}$. For all structures, a strong increase in η_i can be seen with increasing $\Delta\,E/k_B T_{HS}$. Two sections of different behavior can be made out. First, for $x \geq 7.5\%$, η_i follows a universal function of $\Delta\,E/k_B T_{HS}$ for all structures, saturating at $\eta_i = 96\%$. For $x < 7.5\%$, however, η_i differs significantly between structures at fixed $\Delta\,E/k_B T_{HS}$. This means, that η_i is not a sole function of $\Delta\,E/k_B T_{HS}$, but is a function of *both* T_{HS} and ΔE. The two different behaviors can be explained with additional leakage currents due to drift and diffusion as laid out in [Bou93]

$$J_{\text{Leak}} \propto T_{HS}^{n} \exp\left[-\Delta E/k_B T_{HS}\right]. \qquad (5.4)$$

[Bou93] states, that in the case of diffusion dominating the leakage current, the exponent n takes the value of $5/2$, while for a leakage current driven by drift n is $3/2$. In figure 5.11 (a), the differences in η_i between structures for a lower, fixed $\Delta E/k_B T_{HS}$ are consistent with leakage dominating via the T_{HS} ^{n} term. For example, at $\Delta\,E/k_B T_{HS} = 5$, the $x = 0\%$ design is measured at $218\,\text{K}$ and shows $\eta_i = 84\%$, while η_i for the $x = 4\%$ design measured at $288\,\text{K}$ is 59%. At higher $\Delta\,E/k_B T_{HS}$, the asymptotic behavior of the exponential function predominates, suppressing leakage currents. The weak variation of η_i with $\Delta E/k_B T_{HS}$ between structures with $x \geq 7.5\%$ is a sign for this suppressed leakage. The upper limit of $\eta_i = 96\%$ is reached by the $x = 4\%$ structure for the lowest temperature $T_{HS} = 218\,\text{K}$ while the $x = 0\%$ structure is strongly affected by J_{Leak} at all temperatures.

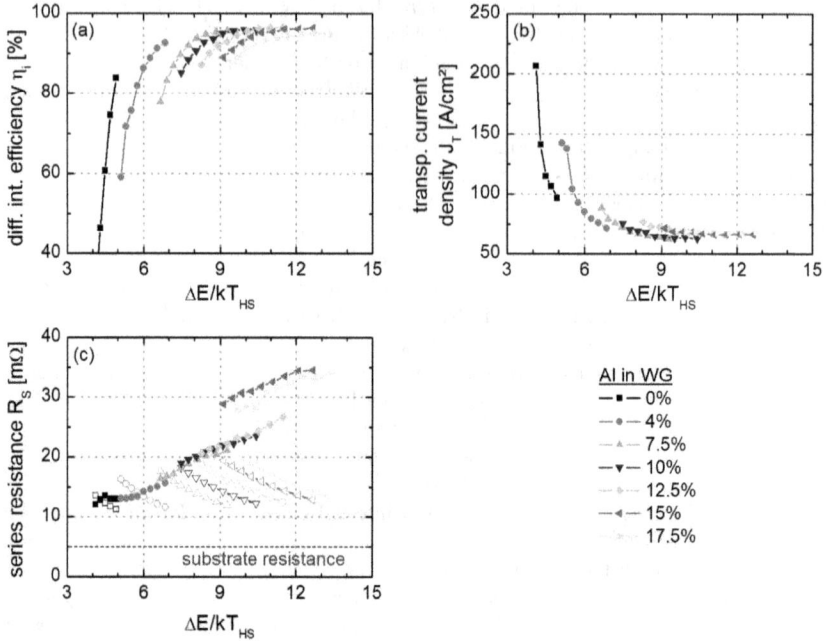

Figure 5.11: Characteristic parameters versus effective barrier height $\Delta E/k_B T_{HS}$ for designs with 0% (squares), 4% (circles), 7.5% (upward triangles), 10% (downward triangles), 12.5% (diamonds), 15% (left triangles) and 17.5% Al-content in the waveguide (right triangles). (a) Calculated differential internal efficiency η_i; (b) calculated transparency current density J_T; (c) measured series resistance R_S and theoretical values calculated by equation 5.1 in open symbols.

Figure 5.11 (b) shows the transparency current density J_T versus the relative barrier height $\Delta E/k_B T_{HS}$. A similar, but inverted behavior to the differential internal efficiency is observed. As expected from the discussion of leakage current, J_T falls approximately exponentially with increasing $\Delta E/k_B T_{HS}$. Again, two sections of different behavior can be made out. For $x \geq 7.5\%$, J_T depends only weakly on $\Delta E/k_B T_{HS}$ and follows a single curve. This is quite different to the behavior of J_T for $x < 7.5\%$, where the transparency current density depends strongly on $\Delta E/k_B T_{HS}$. Also, large differences between different Al-contents, i.e. absolute barrier heights ΔE, are visible. At a value of $\Delta E/k_B T_{HS} \sim 5$, for example, J_T differs significantly between $x = 0\%$ ($96\,\mathrm{A\,cm^{-2}}$ at $218\,\mathrm{K}$) and $x = 4\%$ ($137\,\mathrm{A\,cm^{-2}}$ at $288\,\mathrm{K}$). Thus, J_T depends on both T_{HS} and ΔE in this region. Aligned with the discussion of the internal efficiency above, the two regions are defined by the absence of leakage for high $\Delta E/k_B T_{HS}$ and dominating leakage for small $\Delta E/k_B T_{HS}$.

The measured resistance is plotted in figure 5.11 (c) against the relative barrier height $\Delta E/k_B T_{HS}$ in full symbols. For every device, R_S increases with increasing $\Delta E/k_B T_{HS}$

and appears to follow a universal function. The resistance of the two structures with the highest Al-contents shows an offset to this function, but is still increasing at the same rate with increasing $\Delta E/k_B T_{HS}$.

Futhermore, the expected bulk-layer resistance was calculated for all structures at the measured temperatures using equation 5.1. The temperature dependent electron and hole mobilities were taken from [Sot00] while carrier density was assumed to be constant over temperature. The latter assumption is in accordance with chapter 2 stating that carrier concentration only varies marginally compared to the change in mobility in the temperature range 200 K-300 K. The calculated bulk resistances are plotted for every $\Delta E/k_B T_{HS}$ in open symbols in figure 5.11 (c) and fall strongly with temperature, i.e. increasing $\Delta E/k_B T_{HS}$. The theoretical values for the different structures do not join into a universal function as the measured resistances and little change is seen in the expected resistances at the lowest temperature (e.g. highest $\Delta E/k_B T_{HS}$) for each structure, rising from $10.5\,\text{m}\Omega$ for the pure GaAs structure to $15\,\text{m}\Omega$ for the $x = 17.5\%$ design. This is in strong disagreement with the measurements and the difference between measured and calculated R_S for this lowest temperature increases with increasing Al-content. The excess resistance at low temperature is thus a clear function of x and cannot result from common factors among the designs such as degrading contact or substrate resistance. Instead, the additional resistance must accumulate within the structure, with the deteriorated transport across a hetero-junction being the most likely explanation. The potential hetero-barriers can be identified in figure 5.3 as the junctions between

- n-cladding and substrate

- n-cladding and n-waveguide

- p-cladding and contact

- p-cladding and p-waveguide (as observed in [Lei10] for InP lasers)

- waveguides and the active zone.

On the n-side, the barrier between cladding and the substrate lies in a highly doped area and is thus not prone to cause major contributions to the series resistance even at low temperatures. The second n-side barrier between the cladding and the waveguide is a small, equally high step for all structures and hence does not serve as an explanation for the x-dependence of the resistance. The step between the p-cladding and the contact is the same for all structures and in such not dependent on x. The barrier height of the hetero-junction between p-cladding and p-waveguide shrinks with increasing x and is unlikely to cause an increasing resistance with x. Lastly, the interface between the waveguides and the active zone shows a negative impact on the series resistance. As seen in figure 5.11 (c), the measured resistance follows a universal line with R_S increasing monotonically with increasing barrier height $\Delta E/k_B T_{HS}$. This suggests, that degraded transport into the active region causes the series resistance in excess of the calculated bulk layer resistance. Facilitating the transport could be achieved by introducing a light doping into the undoped active zone (without introducing significantly deteriorating optical absorption) or by changing the transition from waveguide to active region, e.g. by implementing different graded index layers [Mor91].

5.2 High power operation of low series resistance designs

The designs developed and tested in section 5.1 reveal a clear path to high powers and high efficiencies. Low Al-content waveguides have the potential to reduce the series resistance without compromising the optical performance at low temperatures. The most promising structure is identified as the design with $x = 4\%$ aluminum in the waveguide. A full laser process containing low Al-content designs and a further promising structure is realized. The resulting single emitters and bars are tested at various temperatures and compared to previous results.

5.2.1 Selection of promising designs

After the low power analysis of the designs assessed in the SLP, the expected power at high bias can be calculated with equation 2.24 with the knowledge of the characteristic parameters, characteristic temperatures T_0 and T_1 and the thermal resistance R_{th}. Gaining these values at the targeted heatsink temperature of $T_{HS} = 203$ K is exemplarily shown for the structure with $x = 4\%$ aluminum in the waveguide featuring the single quantum well in figure 5.12. The extrapolated efficiency values at the targeted operation power of $P_{op} = 20$ W at $T_{HS} = 203$ K from a $w = 100\,\mu$m wide stripe as well as the values for further promising structures are listed in table 5.5.

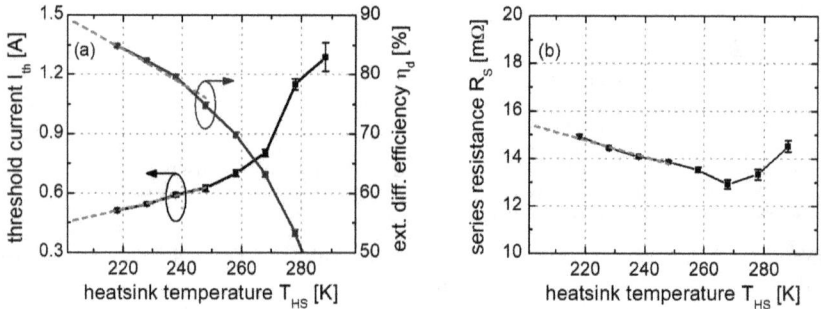

Figure 5.12: Characteristic parameters of the structure with SQW 4% Al-content in the waveguide versus temperature. (a) threshold current and slope efficiency with T_0 and T_1 (dashed) respectively, extended down to targeted operation temperature of $T_{HS} = 203$ K; (b) series resistance with linear extrapolation (dashed) down to 203 K.

In figure 5.12 (a), the measured threshold current and the external differential efficiency for the structure with $x = 4\%$ aluminum in the waveguide and the single quantum well are plotted against the heat sink temperature with the lowest temperature being $T_{HS} = 208$ K. An exponential fit following equations 2.21 and 2.22 is performed in the temperature range $T_{HS} = 218\ldots248$ K to gain T_0 and T_1 (dashed line in figure 5.12 (a)). This resulting exponential fit is extended down to $T_{HS} = 203$ K to determine the values for I_{th} and η_d at this temperature. In figure 5.12 (b), the measured series resistance in plotted against the heatsink temperature. By linear extrapolation of the values in the temperature range $T_{HS} = 218 \ldots 248$ K, R_S at the targeted temperature of $T_{HS} = 203$ K is found. Little to no change is seen in the thermal resistance R_{th} in the temperature range of interest. For example, the conducted CW measurements on the p-side up mounted diodes with the $x = 12.5\%$ design showed $R_{th} = 5.7$ K/W at 298 K and 5.9 K/W at 228 K. Thus, R_{th} is taken to be the same at $T_{HS} = 203$ K as at $T_{HS} = 218$ K. The targeted measurement conditions of 1.2 ms pulses with a frequency of 10 Hz entail $R_{th} \sim 0.9$ K/W at this temperature. Assuming the voltage follows a linear function as in equation 5.3 the current dependent power, voltage and efficiency can be calculated. Minor changes in $U_{d,0}$ do not significantly influence the sought efficiency at 20 W and $U_{d,0}$ is taken from the measurements at $T_{HS} = 218$ K. In table 5.5, the calculated and extrapolated values are listed which are used with equations 2.24 and 5.3 to gain the efficiency values at $P_{op} = 20$ W and the peak efficiency.

Table 5.5: Expected values at $T_{HS} = 203$ K for different structures and the extrapolated peak efficiency and efficiency $\eta_{20\,W}$ at $P_{op} = 20$ W for a $w = 100\,\mu m$, $L = 4\,mm$ stripe.

Structure	I_{th} [mA]	η_d [%]	$U_{d,0}$ [mV]	R_S [mΩ]	T_0 [K]	T_1 [K]	R_{th} [K/W]	η_{peak} [%]	$\eta_{20\,W}$ [%]
Asloc4SQW	461	89	20	15.4	144	305	0.9	74.9	68.2
Asloc7.5SQW	421	91	20	19.7	252	846	0.9	76.4	69
Asloc4DQW	598	87	12	16	179	629	0.9	72.3	67.2

The designs listed are expected to sustain high efficiencies up to the required high powers and are subsequently chosen for a full laser process.

5.2.2 Fabrication of low operation temperature lasers

In a full laser process, the following designs are processed into single emitters and bars:

- **Asloc4SQW**: lowest R_S design due to the low $x = 4\%$ Al-content in the waveguide. Promises to reach the targeted powers of $P = 1.5$ kW with an efficiency close to $\eta_E = 70\%$ in bar format.

- **Asloc7.5SQW**: a more conservative approach with a slightly higher Al-content in the waveguide.

- **Asloc4DQW**: more stable high current operation due to the DQW [Pie09], lower temperature sensitivity.

- **Asloc4DQWpdop** with a lightly **p-doped active region** to assess the benefit to series resistance.

Room temperature length dependent analysis of the Asloc4DQWpdop reveals, that $\alpha_i = 0.96$ cm^{-1} and $J_T = 186$ A/cm^2, both of which are increased compared to the Asloc4DQW structure, while the values of η_i and Γg_0 are unchanged (cf. table 5.3). The lateral layout is slightly changed in comparison to the full process in iteration 0. Since the cost of a diode laser is driven by the material used, bars with a longer resonator are more expensive. The price of the diodes is crucial for their use as pump sources for laser ignited fusion applications [Der11] so shorter cavity lengths are preferred. The layout of the full process takes this into account, limiting the bar resonator length to $L = 4\,$mm and omitting the 6 mm long bars. High fill factors are realized in two configurations:

- FF = 69 %: 37 stripes with $w = 186\,\mu$m (lat. layout A)

- FF = 72 %: 18 stripes with $w = 400\,\mu$m with a structured contact of $5\,\mu$m with $10\,\mu$m pitch to suppress ring lasing and stabilize the near field (lat. layout B) as suggested in [Spr10].

Additionally, single emitters with resonator lengths of 4 mm are processed for comparability with the bars and the previously short-loop processed designs. The facets of the bars and single emitters are passivated for high output powers and subsequently coated for high reflectivity for the rear facet and low reflectivity for the front facet. Single emitters are soldered p-side down onto CuW- heat spreaders and the n-side is connected with bond wires. Laser bars are sandwiched between two CuW heat spreaders using AuSn as solder. Initially, this sandwich is soldered p-side down using PbSn onto a conduction cooled package (CCP) and the n-side is contacted via a 50 µm thick copper foil. This configuration ensures high mounting flexibility due to the copper foil and promises low mounting stress on the bar. However, it is later found that the Cu-foil adds significant series resistance due to the thin conducting cross section. A revised mounting scheme for the laser bars replaces the Cu-foil as the n-side contact by a solid copper contact with a high cross section of 20 mm^2, a fourty fold increase in copper cross section compared to the Cu-foil. Here, the sandwiched bar is soldered between the CCP and the massive n-contact using In which due to its very ductile nature protects the sandwich from mounting and temperature induced stress. Using the thicker n-contact reduces the package resistance of the mounted bar to $\sim 10\ \mu\Omega$, compared to $\sim 200\ \mu\Omega$ for the Cu-foil variant. The absolute numbers were determined from measuring the current-voltage curve from dummy packages with the two mounting schemes.

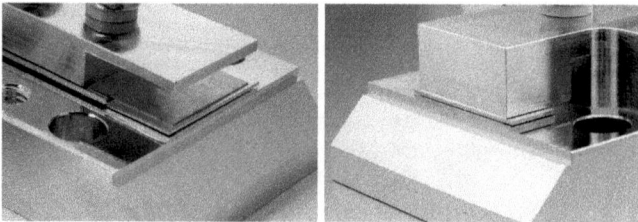

Figure 5.13: Packaged bar using a Cu-foil (left) with a package resistance $\sim 200\ \mu\Omega$ and a massive copper piece (right) for contacting the n-side entailing $\sim 10\ \mu\Omega$.

5.2.3 Characterization of single emitters

As a first quality control, single emitters with $w = 100\,\mu m$ and $L = 4\,mm$ are tested under QCW condition (1.2 ms, 10 Hz) at heatsink temperatures in the range of $T_{HS} = 298\,K$ to 218 K. Their low current performance is compared to the previous CW results from the short-loop processed designs to verify that extrapolated powers and efficiency based on the findings from the short-loop process (SLP) can be also expected from fully processed (FP) devices. The power current characteristics for the three designs are shown in figure 5.14 for the FP devices and the single emitters from the SLP, while the temperature dependence of the series resistance is compared in figure 5.15.

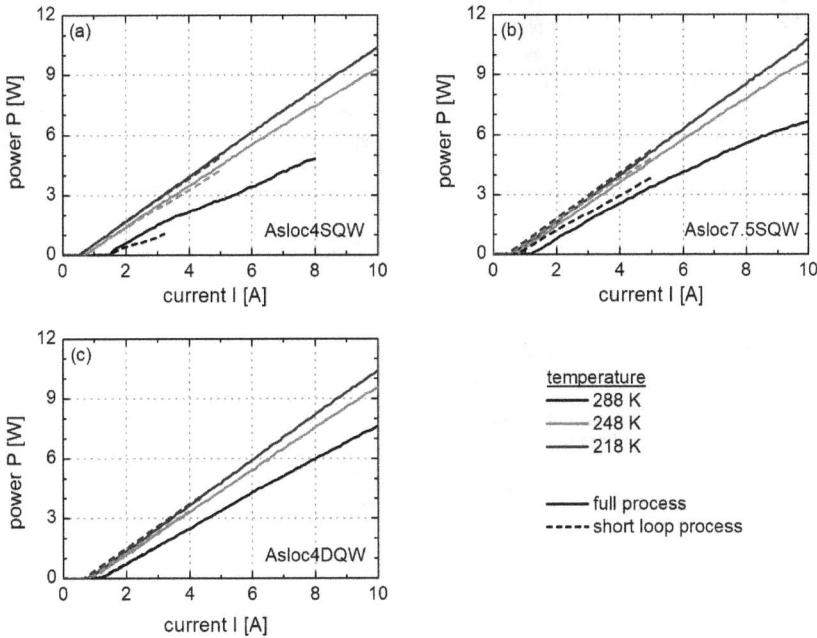

Figure 5.14: Power-current characteristics for $w = 100\,\mu m$ and $L = 4\,mm$ diodes from the full process (solid) and short loop process (dashed) for the designs (a) Asloc4SQW, (b) Asloc7.5SQW and (c) Asloc4DQW.

It is important to note that the thermal load is different between the FP and SLP devices. First, previous measurements on SLP diodes were performed under CW conditions for the most reliable power calibration. With the establishment of the measurement process and calibration improvement QCW measurements are now possible with confident power specification. The higher CW load of the earlier measurements also coincides with a higher thermal resistance of the SLP chips due to the p-up mounting (versus the p-down assembly of the FP devices), resulting in $R_{th} \approx 5 - 6$ K/W compared to 0.9 K/W. Together, this

87

will deteriorate performance at higher temperatures and currents of the SLP devices compared to the FP ones. Nonetheless, comparison of threshold current, resistance and slope at low currents will give a valid indication of previous results.

In figure 5.14 (a) the Asloc4SQW design shows strongly different optical performance for FP and SLP diodes at the highest temperature. Still suffering from roll over, the FP devices have a higher slope and reach significantly higher powers than the SLP chips. Being the most temperature sensitive design, this is easily explained by the higher thermal load of the CW operated SLP diodes. At T_{HS} = 248 K, the slope and threshold current of the FP and SLP devices are very similar. A slightly degraded slope is still visible for the SLP chips. Finally, at the lowest temperature I_{th} and S are identical and only small deviations are seen at the highest comparable currents due to an increased roll over of the SLP diodes. This shows that the previous findings for this vertical design are reproducible in fully processed single emitters and thus bars of the same material.

The Asloc7.5SQW design displays strong differences between the FP and SLP diodes in figure 5.14 (b). Contrary to expectation, FP diodes perform worse at all temperatures due to increased threshold current and degraded slope for the higher temperatures. This is caused by an adjustment of the epitaxial recipe between the SLP and FP. After the short-loop process, samples of the wafer material are routinely analyzed with a scanning electron microscope using backscattered electron imaging to confirm the layer thicknesses. For the samples of the Asloc7.5SQW design, significant deviation of the measured to the nominal layer thickness were discovered as shown in table 5.6.

Table 5.6: Nominal and measured layer thickness of SLP and FP diodes of the Asloc7.5SQW design.

Layer	$d_{nom.}$ [nm]	$d_{SLP,meas.}$ [nm]	$d_{FP,meas.}$ [nm]
p-contact	160	180	160
p-cladding	330	360	340
p-waveguide	975	1170	1040
active zone	7	7	7
n-waveguide	1470	1730	1560
n-cladding	1930	2240	2050

For example, the p-waveguide proved to be 200 nm thicker than targeted. Subsequently the nominal thickness of the layers was adjusted in the epitaxial recipe of the FP diodes to match the measured thickness of the SLP diodes and doping was lowered in the waveguides to match the expected internal loss. Instead, measured thicknesses proved to be closer to the original nominal thicknesses. Furthermore the wavelength is shifted significantly between the FP and SLP diode using the Asloc7.5SQW design to match the targeted wavelength. The SLP diodes lase at λ_c = 958 nm at a temperature of T_{HS} = 218 K and the FP diodes are now centered around 937 nm at this temperature, in accordance with target specifications. Thus, the FP epitaxial design is no longer consistent with the SLP design and together with the lowered doping, results can not be consistently compared to previous conclusions on the Asloc7.5SQW design. This challenge was only present in the Asloc7.5SQW design, as the epitaxial recipes of the other designs remained unchanged and the measured layer thickness between FP and SLP match.

The high temperature stability of the Asloc4DQW design is once more revealed in figure 5.14 (c) as the FP and SLP diodes show the same optical performance at all three temperatures. Even the higher thermal load does not affect the SLP diodes and the vertical design has been successfully reproduced in the FP single emitters and thus eventually bars.

In figure 5.15 the measured resistances of the FP and SLP diodes for the three designs are compared. The diodes with the Asloc4SQW vertical design show good agreement in series resistance for the FP and SLP version with significant differences only at higher temperatures where the design shows poor performance. For the Asloc7.5SQW design the effect of the altered layer thicknesses and the changed doping on the series resistance is clearly visible. The resistance of the FP diodes increases more strongly with decreasing temperature than the SLP chips resulting in a significantly higher R_S at the lowest temperatures. Together with the deteriorated optical performance, this means that the fully processed Asloc7.5SQW design cannot be expected to reach the extrapolated efficiency levels from table 5.5. Finally, there is an absolute offset of 2 - 3 mΩ between the FP and SLP diodes for the Asloc4DQW design, but development of series resistance with temperature is identical.

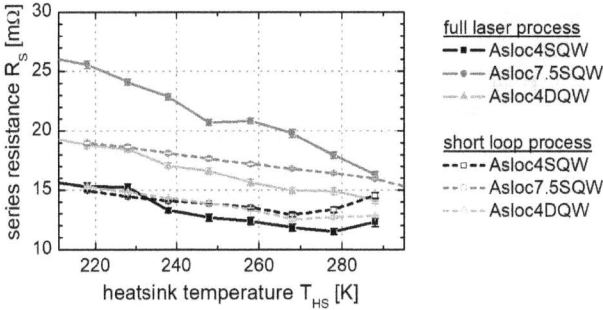

Figure 5.15: Series resistance versus temperature for $w = 100\,\mu$m and $L = 4\,$mm diodes from the full process (solid) and short loop process (dashed) for the designs Asloc4SQW (circles), Asloc7.5SQW (squares) and Asloc4DQW (triangles).

After comparing the low-current measurement results of the FP diodes to the previous analysis of the SLP devices, the FP single emitters are tested to higher currents at $T_{HS} = 208\,$K in QCW mode (1.2 ms, 10 Hz). Measurements up to $I = 20\,$A of the three vertical structures are compared to the $T_0 T_1$ extrapolated light-voltage-current curves (based on the SLP results) in figure 5.16.

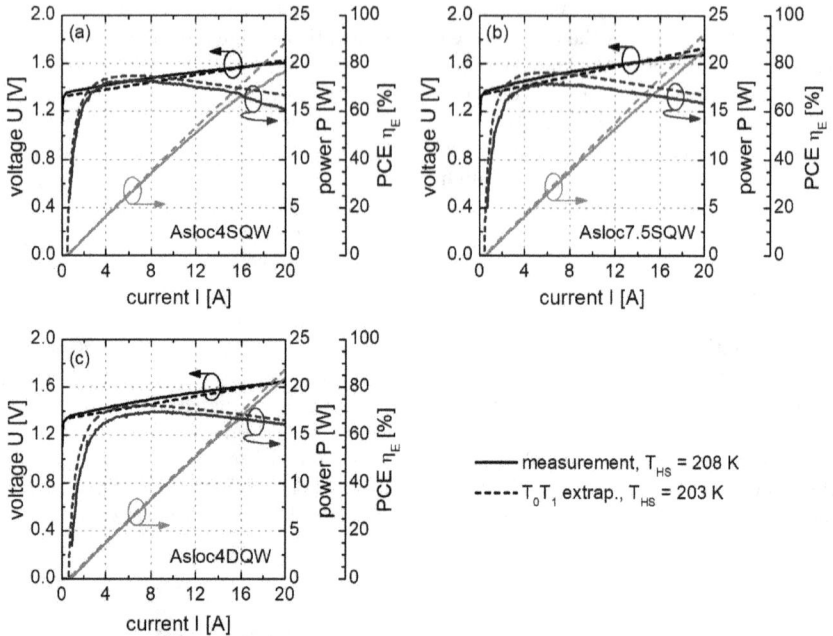

Figure 5.16: Comparison of extrapolated current dependent values of power, voltage and efficiency with measurements for the designs (a) Asloc4SQW, (b) Asloc7.5SQW and (c) Asloc4DQW. Measurements are performed at 208 K under QCW conditions (1.2 ms, 10 Hz) on $w = 100\,\mu m$, $L = 4\,mm$ diodes.

In figure 5.16 (a), the electrooptical performance of the Asloc4SQW design is shown. The light current curve reveals a strong roll over starting at a current of $I = 12\,A$. Simultaneously the voltage current characteristic begins to deviate from its ideal linear form with series resistance decreasing. All in all, the structure reaches high peak efficiencies of $\eta_{peak} = 73\%$ at $I = 6.1\,A$ which is close to the extrapolated values. However, due to the unforeseen roll over the targeted power level of $P_{op} = 20\,W$ is not reached and efficiency is also strongly deteriorated at high currents. The characteristic parameters of the FP diodes are summarized in table 5.7 and extrapolated and measured values are compared.

The roll over is attributed to the accumulation of carriers in the p-waveguide due to thermionic emission and bend bending comparable to reported observations in [Wan10], subsequently studied separately in detail in [Has17]. The higher carrier concentration causes an increase in optical absorption and the internal loss α_i increases along with the threshold current, deteriorating external differential efficiency and power and adding to waste heat which in turn will further increase the leakage. Especially the effect of the tilt of the band edges driving the leakage current at high injection levels is not taken into account in the T_0T_1 extrapolations which focuses on the temperature dependence of threshold current and slope extracted from low current measurements. A possible solution to reduce the leakage is to increase the barriers around the quantum well, as shown in the discussion of $\Delta E/k_B T_{HS}$ in section 5.1.5. The lack of accounting for high current phenomena explains the difference between the extrapolation and measurement.

Table 5.7: Measured and extrapolated values at $T_{HS} = 208$ K and 203 K, respectively, for different vertical structures for a BAL with a $w = 100\,\mu$m, $L = 4$ mm stripe.

Structure	I_{th} [mA]	η_d [%]	$R_S(I_{th})$ [mΩ]	$R_S(20\,A)$ [mΩ]	$P_{meas.}(20\,A)$ [W]	$P_{extr.}(20\,A)$ [W]
Asloc4SQW	515	87.6	15.7	9.1	19.1	22.2
Asloc7.5SQW	547	88.7	26.2	9.6	21.3	23.2
Asloc4DQW	782	87.4	19.5	10	20.9	21.8

Structure	$\eta_{E,meas.}(20\,A)$ [%]	$\eta_{E,extr.}(20\,A)$ [%]	$U_{meas.}(20\,A)$ [V]	$U_{extr.}(20\,A)$ [V]
Asloc4SQW	60.6	66.7	1.61	1.63
Asloc7.5SQW	63.3	66.9	1.68	1.73
Asloc4DQW	64.4	66.1	1.65	1.65

The high power measurement of the Asloc7.5SQW design depicted in figure 5.16 (b) illustrates again the negative effect of the adjusted layer thickness combined with the lower doping. The higher Al-content waveguides lead to an increased series resistance and a slightly higher threshold current, lowering the peak efficiency from the expected $\eta_{peak} = 76.4\%$ to 72.4%. The power level of $P_{op} = 20$ W is reached, but the efficiency is reduced to $\eta_{20\,W} = 64.8\%$. Roll over of the power current characteristic can be observed starting at $I \sim 10$ A, significantly lowering the output power at $I = 20$ A by ~ 2 W compared to the expected value. Again, this unforeseen roll-over is attributed to the high current band bending which was not accessible by the low current measurements used to extract the parameters for the T_0T_1 extrapolation. A further difference between the extrapolation and the measurement pertains to the development of the voltage with current. As mentioned before resistance is higher than expected at low currents. However, a strong bending of the UI curve is observed at higher currents and resistance is strongly reduced. Eventually, this bending leads to a visibly lower voltage of 1.68 V at $I = 20$ A than predicted (1.73 V). However, this lower voltage is not sufficient to compensate for the loss in output power at high currents. All measured values as well as the extrapolated values for power, voltage and efficiency at the maximum current are summarized in table 5.7.

The Asloc4DQW design was included in the full laser process for its expected power stability while providing low series resistance. This proves to be fulfilled looking at figure 5.16 (c). The measured power current characteristic follows the extrapolated T_0T_1 behavior closely with very slight deviation due to roll over visible only above $I = 14$ A.

While the voltage current characteristics differ at low currents due to the higher series resistance of the FP diode, the measured voltage deviates from its expected linear form and at $I = 20$ A the measured and the expected voltage coincide. Together, this means that $\eta_{peak} = 69.8\%$ is lower than expected but $\eta_{20\,W} = 65.1\%$ is close to the extrapolated value. Comparing the high power performance of the Asloc4SQW and Asloc4DQW design, the greater stability of the DQW design is evident at high powers (no roll over).

All three structures exhibit the same behavior in the voltage current characteristic. At higher currents, the voltage curve deviates from its expected linear form and the series resistance is significantly lowered as witnessed by the slope of the curve and seen from the values of series resistance at threshold current $R_S(I_{th})$ and at the maximum current $R_S(20\,A)$. In figure 5.17 the series resistance was calculated for each current and is plotted against the current. Instead of deriving dU/dI, though, R_S was gained by linear fitting the UI-curve over the interval $[I - 0.8\,A, I + 0.8\,A]$ for smoothing.

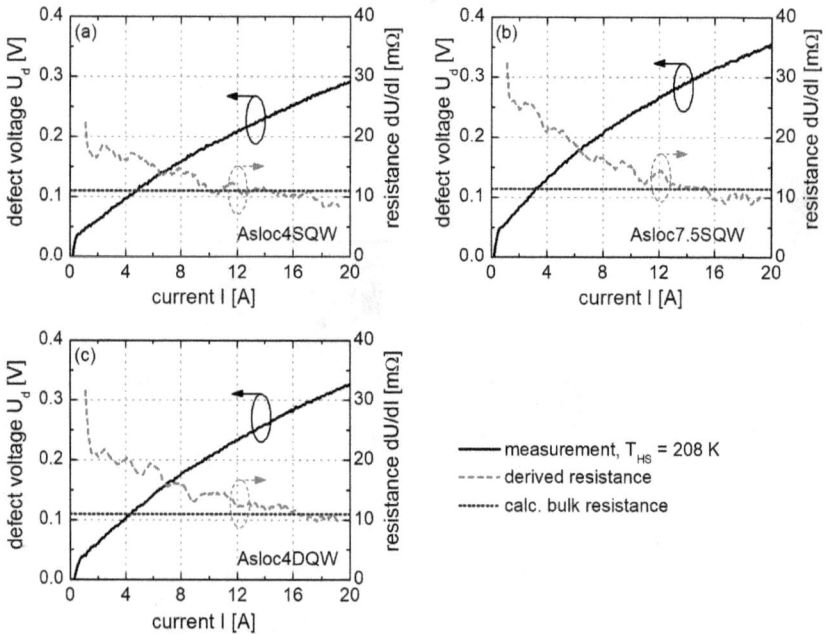

Figure 5.17: Measured voltage (solid) at 208 K in QCW mode (1.2 ms, 10 Hz) and series resistance (dashed) as calculated from fitting the UI-curves versus current for the designs (a) Asloc4SQW; (b) Asloc7.5SQW and (c) Asloc4DQW.

For all three vertical designs two different regions of series resistance are visible in figure 5.17. At low currents, R_S is much increased, matching the observed series resistance from the short-loop processed devices. As current increases, the resistance drops more rapidly for the Asloc7.5SQW design. Starting at a current of around 10 A, R_S decreases at a lesser rate for all structures and little difference in magnitude of R_S is seen between the structures. This high current regime was inaccessible by the short-loop processed devices and measured series resistance here coincides with the calculated bulk resistance from equation 5.1. It appears that the increase in series resistance with barrier height is a low current phenomenon which is overcome at higher current densities at which point the diodes behave in accordance with bulk layer estimation. However, the defect voltage U_d accumulated in the low current regime due to the higher series resistance is much bigger for higher Al-content designs. Thus, the benefit of a low-Al content in the waveguide on the voltage remains though does not extend into the high current regime. For example, the difference in U_d between the Asloc7.5SQW and the Asloc4SQW design at 10 A is 53 mV which does not increase much to 20 A at which point it amounts to 64 mV.

All in all, the high power measurements of single emitters from the designs in iteration 1 prove the benefit of low Al-content waveguides for the efficiency at the targeted high powers. Extrapolating from the single emitter to bar performance the Asloc4DQW design is expected to reach the highest efficiencies at high powers, while the Asloc4SQW design will show the highest peak efficiencies at low currents but suffer from roll over at higher currents.

5.2.4 Characterization of bars

After the benchmarking of single emitters, 1 cm wide high fill factor bars with the different designs were assembled to confirm the good performance seen in the single emitter measurements. Coated bars with lateral layouts A and B were sandwiched between CuW heat spreaders using AuSn and mounted p-side down onto a conduction cooled package (CCP). The n-side was in all cases established by a massive contact for a low package resistance of $R_p \approx 10\mu\Omega$. Precise mounting of the bars is crucial to ensure homogeneous contact with all emitters, leading to uniform pumping and heat removal across the bar. At the same time, mechanical stress from the mounting and soldering process due to different thermal expansion coefficients of the involved materials can cause deteriorated performance of stripes on the bar as examined in [Bul05]. These challenges are carefully addressed in the assembly process. The quality of the mounting is checked by measuring the level of uniformity of the power distribution across the bar. The tests are performed at room temperature under QCW condition (1.2 ms, 10 Hz), the pump conditions for Yb:YAG solid-state lasers. Only low currents are applied to minimize the risk of damaging the device using the measurement setup described in 3. Bars with uniform power distribution are considered well mounted and are consequently tested to high currents at low temperatures.

The two different lateral bar layouts are assessed for the Asloc4SQW design. Nominally, the bars have a similar fill factor of 69% (37 x 186 µm) and 72% (18 x 400 µm). The assembled bars were tested at room temperature up to low currents in QCW mode (1.2 ms, 10 Hz). In figure 5.18 the defect voltage of two bars with the different lateral layouts is plotted against the current. While the higher FF of the bar with motif B is expected to cause a 4% lower series resistance due to the higher contact area, the measurement points into the opposite direction with resistance of the bar with motif B being 0.44 mΩ versus 0.28 mΩ for motif A. The higher series resistance in the case of the higher fill factor is attributed to the structuring of the contact. While the pumped width is 400 µm wide, the width of the contact is reduced to 200 µm due to the periodic structure of 5 µm metal with 10 µm pitch. The current will spread under the slim contact stripes and pump the whole 400 µm stripe. However, it has to pass a much reduced contact area and by equation 5.1, series resistance is increased. This feature had been pointed out in chapter 4 but a thorough comparison of the two lateral layouts had not been performed due to the missing bars with motif A. The higher R_S of motif B means that the lateral motif A promises higher efficiencies at highest powers and all future bars will be of this lateral design.

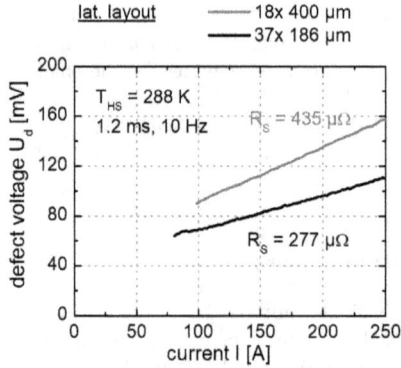

Figure 5.18: Defect voltage versus current for bars with the Asloc4SQW design using the two lateral layouts of 18 x 400 µm and 37 x 186 µm.

The initial mounting scheme featured a thin copper foil as the n-contact which had been chosen for flexibility and ease of assembly. Bars with the vertical design Asloc4SQW and the lateral layout with 37x 186 µm stripes using this contact show a strong degradation in efficiency at higher currents. The comparison to the same bars mounted with the revised mounting scheme (massive n-contact) measured at T_{HS} = 203 K in QCW mode (200 µs, 10 Hz) is shown in figure 5.19 (a). The lower pulse width was chosen to decrease the duty cycle and thus the dissipated heat, as the lower limit of operation temperature was reached for the measurement setup.

Looking at the power, voltage and efficiency characteristics in figure 5.19 (a), the difference in n-side contact is clearly visible. While the threshold current and slope coincide, the bar with the Cu-foil experiences a stronger roll over in power current characteristics, only reaching output powers of 1.55 kW compared to the 1.69 kW of the massive n-side contact version. However, this difference can not be explained by the different n-side contact (for example due to ohmic heating of the copper foil) and is attributed to deviation from bar to bar. The difference in voltage current characteristic is significant, though. The copper foil adds a package resistance of ~ 0.2 mΩ which is in the same order of magnitude as the resistance from the semiconductor material itself. Voltage is clearly increased and thus the efficiency is considerably reduced. At a power of 1.5 kW, the efficiency of the bar with the massive n-side contact is η_E = 55% (with a peak efficiency of 69%) and is

reduced to 39% for the bar contacted by the Cu-foil. Hence, the benefit of the thicker n-side contact is proven and it was used in all future bar assemblies.
In figure 5.19 (b) the integrated spectrum at the different currents is displayed for the bar with the massive n-side contact measured at T_{HS} = 203 K in QCW mode (200 µs, 10 Hz). Starting at λ = 934.4 nm right above threshold, the wavelength shifts with increasing current to the targeted λ = 939.3 nm at the required current for 1.5 kW output power.

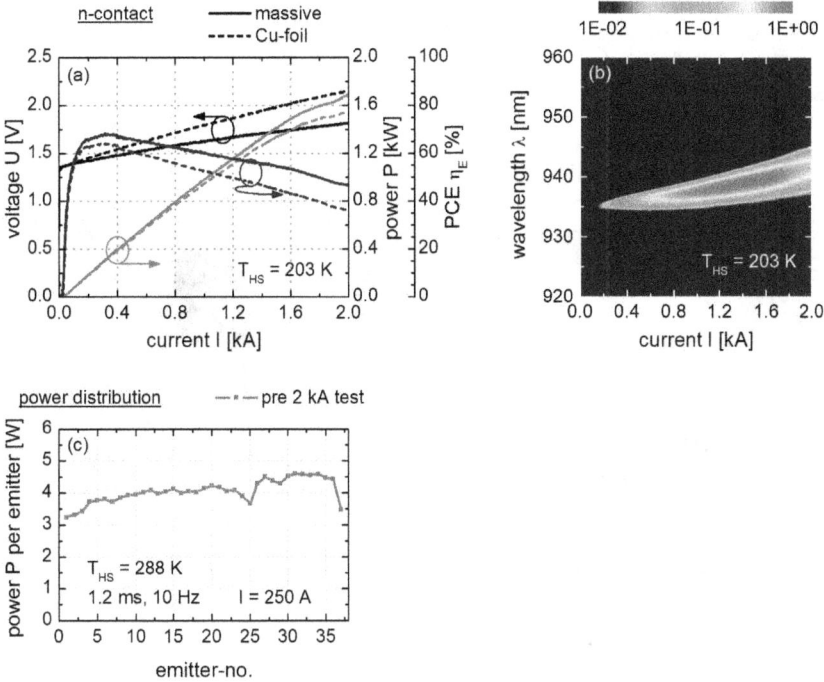

Figure 5.19: Measurements of the Asloc4SQW design in bar format. (a) Power, voltage and efficiency versus current for the two n-side contacts Cu-foil (dashed) and massive (solid); (b) integrated spectrum of the bar with the massive n-side contact; (c) power distribution across the bar with the massive n-side contact prior to 2 kA testing.

Prior to testing the bar to higher currents at lower temperature, a histogram across the emitters of the bar was recorded at a current of 250 A at room temperature in QCW mode (1.2 ms, 10 Hz), as shown in figure 5.19 (c) for the bar with the massive n-side contact. The power of each single emitter was measured and the power distribution across the bar is plotted. The power varies between 3.2 W and 4.5 W with a standard deviation of the mean of σ_P = 0.36 W. In accordance with [Cru04] the inhomogeneity of emitted power across the bar is attributed to mechanical stress within the active region due to the assembly, although non-uniform current injection and degraded emitters on the bar also affect the emitter power homogeneity across the bar.

The Asloc4SQW design measured in bar format confirms the conclusions drawn from the measurement of single emitters with this design: even at lowest temperatures, significant roll over of the optical output power is observed at higher currents. While efficiency is increased at low currents due to the low series resistance, the loss of output power at high currents overcompensates the benefit of the low R_S. Furthermore, non-ideal emitter yield and stress effects in the bar introduce additional deterioration to power, as seen in the histogram. Subsequently, further bars with the other vertical designs from the FP were assembled and tested to high currents at low temperatures. The lessons drawn from the measured bars with the Asloc4SQW design were applied and only bars with the lateral layout comprised of 37x 186 μm stripes were assembled using a massive n-contact. In figure 5.20 the results of the Asloc7.5SQW in bar format are shown.

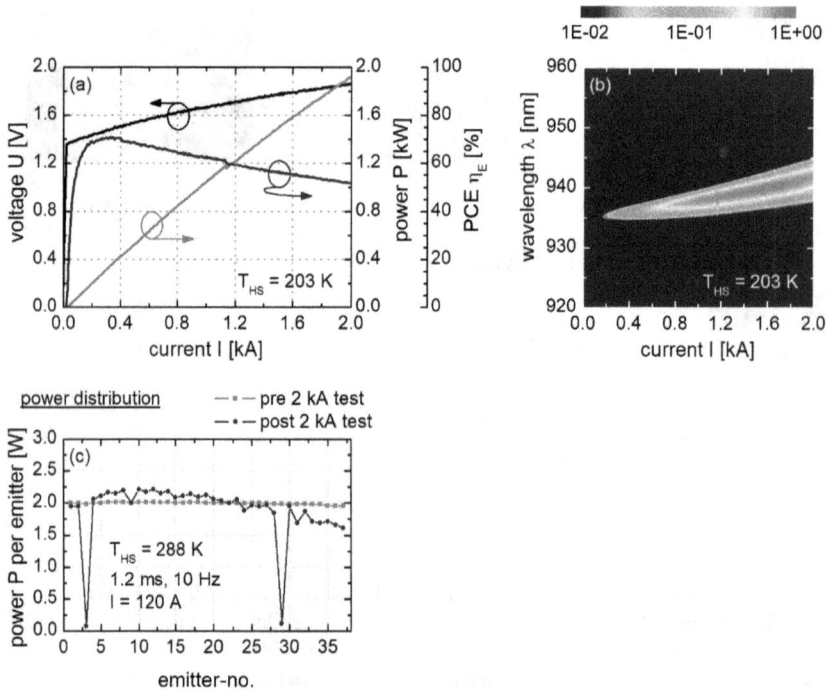

Figure 5.20: Measurements of the Asloc7.5SQW design in bar format. (a) Power, voltage and efficiency versus current; (b) integrated spectrum; (c) power distribution across the bar prior (squares) to and after (circles) 2 kA testing.

The electro-optical performance in QCW mode (200 μs, 10 Hz) at 203 K is depicted in figure 5.20 (a). At 1.5 kW the bar achieves an efficiency of 56% reaching 1.92 kW at 2 kA with an efficiency of 52%. Peak efficiency is 71%, dropping rapidly with increasing current due to a much higher series resistance compared to the bar with the Asloc4SQW design in agreement with the prior observations on single emitter level. Roll over of power is much reduced compared to the Asloc4SQW design, but voltage at 2 kA is in turn 40 mV higher (1.86V). Furthermore, two kinks are seen in the light current curve at 0.4 kA and 1.13 kA. As no sudden spectral shift is seen at these points (figure 5.20 (b)), the kinks are attributed to emitter failures.

The recorded spectrum is shown in figure 5.20 (b), measured under the same conditions. Above threshold, the wavelength is centered around $\lambda = 934.7$ nm, shifting and broadening with increasing current. At the current needed to achieve 1.5 kW in optical power, the central wavelength is at 939.2 nm, ideal for pumping Yb:YAG solid-state lasers. Comparing the spectra of the Asloc7.5SQW to the Asloc4SQW design, the stronger power roll over of the latter becomes visible, as the spectrum widens more strongly at the onset of the roll over. At 1.5 kW, spectral width with 95% power content is $\Delta\lambda_{95\%} = 8.3$ nm for the Asloc4SQW bar compared to 7.3 nm for the Asloc7.5SQW bar. The distribution of power across the bar was measured prior and after the test up to 2 kA and is shown in figure 5.20 (c). The histograms were recorded at room temperature at low currents of 120 A in QCW mode (1.2 ms, 10 Hz), confirming the failure of two emitters. While the power was homogeneously distributed across the bar with an average power of $P = 2$ W prior to high current testing and with a standard deviation of the mean of $\sigma_P = 0.02$ W, significant differences in output power amongst the emitters are observed after. Omitting the two failed emitters, the standard deviation of the mean rose to $\sigma_P = 0.18$ W.

The measured bar values of the two designs are compared in table 5.8. Overall, the bar with the Asloc7.5SQW design achieves higher output powers at higher efficiencies than the one with the Asloc4SQW design. The maximum output power is decreased by the failure of two of the 37 emitters while roll over at higher currents is less pronounced than for the Asloc4SQW case. However, the significantly increased voltage due to the high series resistance considerably reduces the efficiency at the high powers.

Measurements for a bar with the Asloc4DQW design are shown in figure 5.21. As previously discussed, the bar contained 37 emitters with a stripe width of 186 μm and was contacted using a massive n-contact.

In figure 5.21 (a) the electro-optical performance of the bar is displayed when measured in QCW mode (200 μs, 10 Hz) at 203 K. No kinks can be seen in the power-current characteristic as well as little roll over at high currents. The bar reaches an output power of $P = 2$ kW at the maximum available current of $I = 2$ kA, the highest yet reported power from a 1 cm bar [Fre15]. The structure benefits from the low Al-content in the waveguide, keeping resistance low and limiting voltage at 2 kA to $U = 1.81$ V. Peak efficiency is $\eta_{E,max} = 69\%$, while the lower series resistance and the reduced roll over lead to high efficiencies at high power, namely 61% at 1.5 kW and even 55% at the maximum power of 2 kW.

The development of the wavelength with increasing current measured under the same conditions is shown in figure 5.21 (b). At the onset of lasing, the central wavelength is at $\lambda = 934.8$ nm, increasing with higher current. At $I = 1.44$ kA at which current the device reaches $P = 1.5$ kW, the central wavelength is shifted to 940.5 nm with a spectral width with 95% power content of $\Delta\lambda_{95\%} = 7$ nm. Finally, at the maximum current, the spectrum is centered at 943.2 nm with $\Delta\lambda_{95\%} = 10$ nm. Prior to subjecting the bar to low temperatures and high currents, it was tested at 288 K at a current of 120 A in QCW mode (1.2 ms, 10 Hz) and the power emitted by each emitter was recorded. This measurement was repeated after the test up to 2 kA and the results are depicted in figure 5.21 (c). Prior to high current testing, the bar showed an average power per emitter of $P = 1.37$ W with a standard deviation across the bar of $\sigma_P = 0.08$ W. The uniformity is thus better than the bar with the Asloc4SQW design, but significantly worse than the Asloc7p5SQW design bar. After reaching 2 kW, the bar has lost no emitters but uniformity is deteriorated with a standard deviation of $\sigma_P = 0.17$ W.

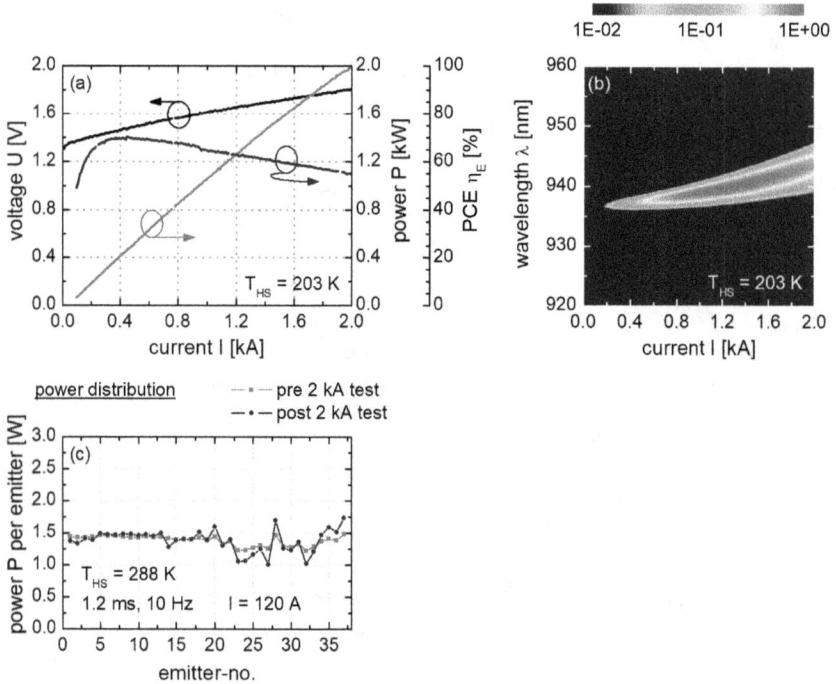

Figure 5.21: Measurements of the Asloc4DQW design in bar format. (a) Power, voltage and efficiency versus current; (b) integrated spectrum; (c) power distribution across the bar prior (squares) to and after (circles) 2 kA testing.

Collectively, the bar using the Asloc4DQW design achieves the highest output power ever reported from a single bar of 2 kW at a very high efficiency of 55%. This is facilitated by the use of the double quantum well, suppressing the roll over of output power at high currents and the use of a low Al-content waveguide to reduce series resistance.

Finally, the results from the Asloc4DQWpdop design in bar format are compared to the best performing Asloc4DQW design in figure 5.22. The designs are compared in their electro-optical performance up to currents of $I = 1$ kA to portray the benefit of using a lightly doped active region. At higher currents, the few available samples from the Asloc4DQWpdop design suffered from emitter failures, not reaching $I = 2$ kA. The current dependent measured output power is displayed in figure 5.22 (a). Measurements were performed at 203 K in QCW mode (200 µs, 10 Hz). Up to a current of $I = 1$ kA, no difference in output power is visible between the doped and undoped version. This shows, that the light doping of $p = 5 \times 10^{-16}$cm^{-3} of the active region does not lead to a significantly lower slope and increased threshold current, despite the higher internal loss and transparency current density seen from L-dependent room-temperature measurements. However, at a current of 0.95 kA one of the emitters on the bar with the Asloc4DQWpdop fails, manifesting in a kink in the power current characteristics and no comparison to higher currents is performed.

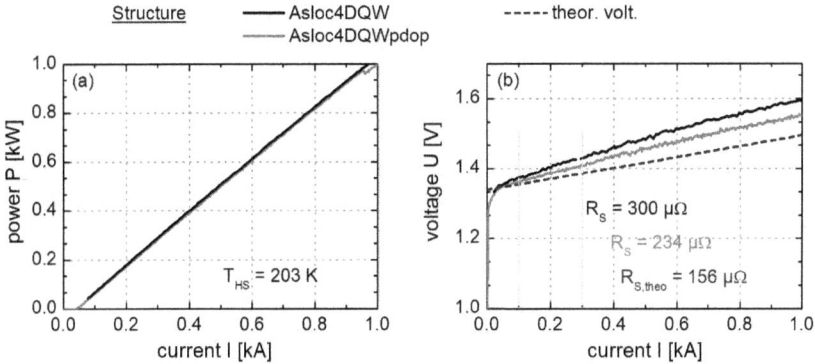

Figure 5.22: Comparison of the Asloc4DQWpdop design to the Asloc4DQW in bar format. (a) Power versus current; (b) voltage versus current. In dashed blue is shown the theoretical current voltage characteristic with the calculated bulk layer resistance. The fit interval is shown in vertical turquoise lines.

In figure 5.22 (b) the voltage current characteristics of the two designs in bar format are shown. Here, a substantial benefit of doping the active region is evident as the resistance is decreased from $R_S = 0.30$ mΩ to 0.23 mΩ. This results in a decrease in voltage at $I = 1$ kA from $U = 1.6$ V to 1.5 V and is expected to enhance efficiency in bars sustaining currents up to 2 kA. At 1 kA, the difference in series resistance has decreased to 0.01 mΩ due to the stronger bending of the UI-curve of the Asloc4DQW structure. Assuming no power degradation due to the doping of the active region and estimating a reduction in voltage between 0 kA and 2 kA of $\Delta U = 0.11$ V, a 4% point increase in efficiency is expected at 2 kW.

The simple change of the light doping proves that the properties of the active region are in fact responsible for the excess resistance beyond the bulk layer contribution which is also shown in figure 5.22 (b) at 0.16 mΩ. This supports the thesis, that the degraded transport of carriers into the active region is responsible for the excess resistance. Introducing the light p-doping has eliminated 50% of the excess resistance without compromising output power and invites further studies to address the excess resistance from the active region, e.g. to find the maximum doping level without deteriorating the slope due to increased optical absorption.

All the values determined from the bar measurements of devices assembled with a massive n-contact are summarized in table 5.8 and compared to the bar results from chapter 4, in which the Cu-foil was still used to contact the chip's n-side. Furthermore, the lateral layout between the bars from Iteration 1 and Iteration 2 is different, as well as the operating temperature.

The measured values of the three vertical designs in bar format are compared in table 5.8.

Table 5.8: Summarized values from QCW bar measurements for different vertical structures from Iteration 1 and Iteration 0. The different n-contact, lateral layout and operating temperature between Iteration 1 and Iteration 0 has to be taken into consideration.

Structure	I_{th} [A]	η_d [%]	$R_S(I_{th})$ [$\mu\Omega$]	$R_S(2\,kA)$ [$\mu\Omega$]	$P_{max.}$ [kW]	$U(2\,kA)$ [V]	$\lambda(1.5\,kW)$ [nm]
Iteration 1 (203 K)							
Asloc4SQW	26.8	81.3	280	170	1.69	1.82	939.3
Asloc7.5SQW	27.2	86	340	170	1.92	1.86	939.2
Asloc4DQW	38.2	84.8	300	180	2	1.81	940.5
Asloc4DQWpdop	43.2	83.3	230				
Iteration 1 (223 K)							
Design2	35.6	70	690	560	1.45	2.55	

Structure	$\eta_{E,max.}$ [%]	$\eta_E(1.5\,kW)$ [%]	$\eta_E(P_{max.})$ [%]
Iteration 1 (203 K)			
Asloc4SQW	69	55	47
Asloc7.5SQW	71	56	52
Asloc4DQW	69	61	55
Iteration 1 (223 K)			
Design2	53		30

The measured performance of the various structures on bar level is compared to the previous results from single emitters using these designs. This analysis sheds further light on possible problems with bar assembly, serves as a quality check for the performance of the bars and give an upper limit for the performance of the designs on bar level. The single emitter results from section 5.2.3 are scaled using the fill factor of the bar. Specifically, drive current and output power are multiplied by 69 while voltage should be equal for bars and single emitters. It should be noted, however, that measurement conditions between single emitters and bars differ slightly in temperature and duty cycle. While single emitters were tested at $T_{HS} = 208\,K$ with 1.2 ms, 10 Hz pulses, bars were operated at 203 K with 200 μs, 10 Hz. Based on the observed temperature dependence of the characteristic parameters of the most temperature sensitive vertical design in figure 5.12, the small difference in temperature is not expected to significantly alter the performance. The impact of the change in duty cycle was assessed by tracking the change in wavelength with power for the single emitters and bars and calculating their thermal resistance, which are listed in table 5.9.

The differences in R_{th} are attributed to the different mounting of the diodes. While the CuW-sandwiched bars are directly soldered onto the CCP using In and are also contacted by the solid Cu-block from the n-side, the SSM with the mounted SE are only pressed down onto the base of the measurement bracket. Furthermore, the individual SSM show different bending of the bottom heatsink surface. A nonuniform higher thermal resistance is thus expected from the single emitters. For the Asloc4SQW design, in which the highest difference R_{th} between SE and bars is measured, the difference in active zone temperature between bar and SE can be calculated with equation 2.23. At a power of 1.2 kW with an assumed efficiency of 60%, the active zone of the SE will be $\Delta T_{AZ,SE} - \Delta T_{AZ,bar} < 6\,K$ hotter and the same argument made for the different heatsink temperatures is made.

Table 5.9: Thermal resistance for bars and scaled single emitters of the vertical structures at T_{HS} = 203 K and 208 K, respectively.

Structure	$R_{th,bar}$ [K/W]	$R_{th,SE}$ [K/W]
Asloc4SQW	0.009	0.016
Asloc7.5SQW	0.01	0.014
Asloc4DQW	0.013	0.015

The scaled single emitter results for the three structures are shown next to the bar results in figure 5.23. In figure 5.23 (a), the electro-optical performance of the Asloc4SQW design in bar and single emitter format is compared. Threshold current and slope coincide, but the bar suffers from an earlier roll over as its slope starts to deviate from the single emitter at $I \sim 0.2\,kA$. At a current of $I \sim 0.9\,kA$, the single emitter is experiencing strong roll over and at the maximum current of $I = 1.38\,kA$ (equivalent to 20 A on single emitter level) the bar and the scaled single emitter emit the same power. The voltage current curve reveals a visible difference between the single emitter and bar. While both approximately start at the same U_0, the bar has a 0.03 mΩ higher series resistance than the single emitter. Together, the lower voltage and reduced roll over lead to a higher conversion efficiency for the single emitter than for the bar. Thus, for an ideal bar, efficiency is expected be $\sim 5\%$ points higher. Generally, no indications are evident that the Asloc4SQW bar suffers from any assembly problems.

The Asloc7.5SQW single emitter and bar are compared in figure 5.23 (b). Similar to the first design, the single emitter shows the same threshold current and slope as the bar. The bar shows an earlier onset of roll over at $I \sim 0.4\,kA$ and reaches a lower output power at the maximum single emitter current. The voltage of the single emitter coincides well with the bar voltage, but a difference in series resistance is seen for currents above 0.4 kA at which point the single emitter shows a lower voltage. This point falls together with the kink in the PI-curve, where a single emitter of the bar failed. Below the current at which the emitter failed, the single emitter and bar data for the Asloc7.5SQW design match, attesting the quality of the bar assembly. In case of no package resistance present and a perfect optical performance of all emitters on the bar, the Asloc7.5SQW design would achieve a $\sim 5\%$ point higher efficiency.

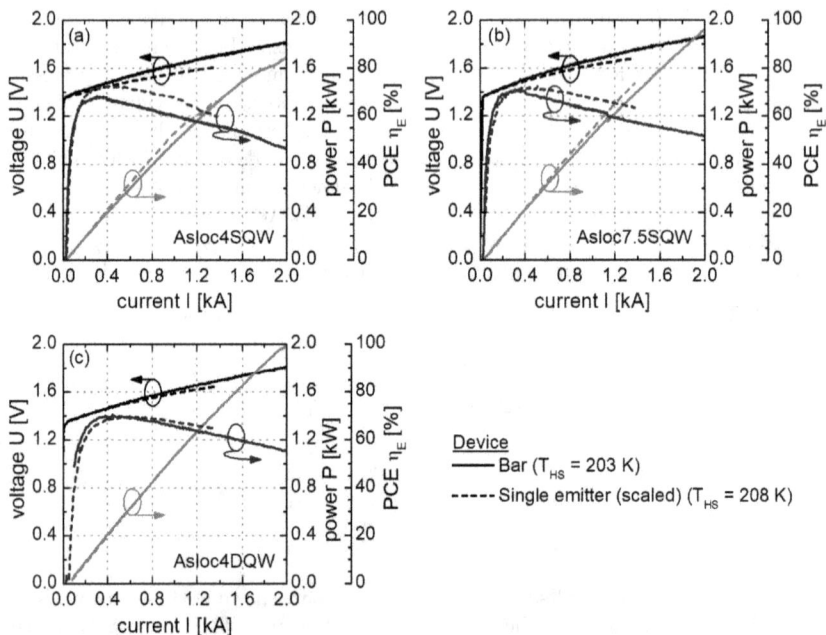

Figure 5.23: Comparison of the single emitter to the bar performance for the (a) Asloc4SQW, (b) Asloc7.5SQW and (c) Asloc4DQW design.

Finally, the measurement results of best performing structure Asloc4DQW in single emitter and bar format is displayed in figure 5.23 (c). No difference can be seen between scaled single emitter and bar in output power as threshold current, slope as well as power roll over are identical. Only minor differences in series resistance are seen at currents above $0.5\,kA$ with the single emitter voltage being lower. Together, this leads to a very good agreement in efficiency of the single emitter and the bar, confirming the ideal performance of the high FF bar.

5.2.5 Estimation of loss mechanisms

In order to further improve the efficiency at high powers, it is important to reduce the waste heat, e.g. the electrical energy not transformed into optical energy P_{opt}. At a given current I, the dissipated power is given by equation 2.19 which can be further expanded to

$$P_{waste}(I) = U(I) \times I - P_{opt}(I) = (1 - \eta_E(I)) \times U(I) \times I. \tag{5.5}$$

The waste heat can further be allocated to different loss channels within the diode. An approach to optimize for higher efficiency is to identifying the main contributors to the dissipated power and reduce its magnitude.

The total waste heat at a current I consists of contributions to the following loss channels:

- establishing threshold
$$P_{I_{th}} = I_{th} \times U(I_{th})$$

- non ideal slope
$$P_{\eta_d} = \left[\frac{hc}{q\lambda} - S\right] \times [I - I_{th}]$$

- power roll over
$$P_{T_0 T_1} = S \times [I - I_{th}] - P_{opt}(I)$$

- defect voltage due to barrier
$$P_{U_{def}} = U_{def,0} \times I$$

- Joule heating due to series resistance or Ohmic loss
$$P_{joule,lin} = R_S \times I^2$$

The Joule heating assumes an ideal, linear UI characteristics and a constant series resistance found from the low current linear fit up to the highest currents. It does not account for the bending of the UI curve described in section 5.2.3 in which a lower resistance at high currents was found. Depending on the linearity of the UI curve up to the assessed current, $P_{joule,lin}$ can thus be severely overestimating the contribution of Ohmic heating to dissipated power. A second option to obtain the Joule heating is by using only the differences between the extrapolated turn on voltage U_0 and the voltage at the maximum current $U(I_{max})$ which is caused by a non specified series resistance, which is not necessarily linear:

$$P_{joule,\Delta U} = [U(I) - U_0] \times I \tag{5.6}$$

This consideration of loss mechanisms is difficult to apply to bar measurements, as failing or non-ideal emitters on the bar will contribute to P_{η_d} or $P_{T_0 T_1}$, and will thus impede conclusions about potential improvements to the vertical structure. Rather, the high power single emitter measurements at $T_{HS} = 208$ K are used to analyze the main efficiency

103

limiting factors by estimating the loss mechanisms at this temperature. It is important to note, however, that some of the contributions might be overestimated by using the existing single emitters, as the narrower stripe width of 100 µm (compared to the 186 µm wide emitters on the bars) is more affected by lateral current spreading, as discussed in section 4.2.2. In the wider emitters used, the waste heat due to $P_{I_{th}}$ and P_{η_d} will be lower. As a point of operation, the highest power reached by all single emitters of the three vertical structures is chosen ($P = 19\,\text{W}$) which corresponds to 1.3 kW from a bar with a fill factor of 69%. The efficiency, total waste heat and the contribution of the individual loss channels at this point are summarized for the three designs in table 5.10. Here, both $P_{\text{joule,lin}}$ as well as $P_{\text{joule},\Delta U}$ are calculated as to illuminate again the strength of the bending of the UI-curve, e.g. the non-uniform series resistance up the high currents.

Table 5.10: Distribution of dissipated power at an optical output power of 19 W across loss channels for the three vertical designs, measured at $T_{\text{HS}} = 208\,\text{K}$ in single emitter configuration with 100 µm stripe width.

	Asloc4SQW	Asloc7.5SQW	Asloc4DQW
η_E [%]	61.5	65.2	65.6
P_{waste} [W]	12.2	10.0	10.2
$P_{I_{th}}$ [W]	0.7	0.7	1.0
P_{η_d} [W]	3.0	2.4	2.7
$P_{T_0 T_1}$ [W]	3.2	1.2	1.2
$P_{U_{\text{def}}}$ [W]	0.5	0.4	0.4
$P_{\text{joule,lin}}$ [W]	6.4	8.3	6.4
$P_{\text{joule},\Delta U}$ [W]	4.9	5.3	5.0

For all structures, the Ohmic loss is the strongest contribution to waste heat, contributing between 40% and 52% of the total distributed power in the case of $P_{\text{joule},\Delta U}$. Due to the decreasing resistance with increasing currents, the expected Joule heating deduced from low series resistance is exaggerated by 30% for the $x = 4\%$ and by 55% for the $x = 7.5\%$ case. Thus, the roll over of the UI-curve benefits the efficiency at high currents and the findings of section 5.2.3 of stronger bending with higher Al-content structure manifest in a lower than expected Joule heating. The second most important loss mechanism is the imperfect slope efficiency, contributing around 25% of the total waste heat in all cases. The slope efficiency is degraded by the internal loss as well as an imperfect internal differential efficiency. The latter is improved by using wider stripes which are implemented in the bars' lateral layout. It is noteworthy, that despite the use of a DQW and the higher internal loss due to free carrier absorption in the active zone, the Asloc4DQW design only shows a minimally higher penalty to P_{η_d} (26% of P_{waste}) than the SQW designs (with 24% of the respective P_{waste}). A possible explanation is that the higher α_i is countered by a higher η_i by suppressed leakage currents. This would explain, why the Asloc4SQW has the highest values for P_{η_d} (highest leakage, low internal loss) and the Asloc7.5SQW the lowest (low leakage due to higher barriers, low internal loss). The last significant contribution to the dissipated power is the power saturation. As already seen from figure 5.16, the Asloc4SQW design suffers strongly from roll over, which amounts to 26% of all waste heat. For the other two vertical designs, $P_{T_0 T_1}$ makes up 11% of the dissipated power. The loss channels $P_{I_{th}}$ and $P_{U_{def}}$ are both below 10% of total waste heat for all designs and are thus not the primary concern for increasing efficiency at the high powers. Even in the case of using a DQW, the elevated values of power needed to establish threshold are negligible.

Generally, the values in table 5.10 give a clear indication on the optimization approach for higher efficiencies at high powers. First and foremost, the power lost to Joule heating has to be reduced without deteriorating any of the other loss channels. Thus, vertical structures with even lower series resistance have to be designed. However, a further reduction in Al-content is not feasible, as even the Asloc4SQW design already suffers heavily from power saturation attributed to the low barrier height around the active region. Increasing the doping will further boost P_{η_d}, which is already the second biggest loss mechanism. Hence, the thickness of the waveguide layers will have to be decreased in future design iterations. A further option is to increase the asymmetry of the waveguide which is proven to reduce both series resistance as well as internal loss [Has14], tackling the two biggest loss channels simultaneously.

5.3 Conclusions from Iteration 1

The approach to reduce the series resistance by lowering the Al-content in the waveguide was explored in iteration 1 and proved to be successful. The first demonstration of 2 kW output power at T_{HS} = 203 K from a single 1 cm bar was presented, which represents a 33% power increase to the targeted output power of 1.5 kW. Low Al-content waveguides ensured high efficiencies of $\eta_E \geq 55\%$ up to the highest powers which was also facilitated by the use of a novel mounting scheme of the bars, reducing package resistance. Compared to the best previously published results [Li08], output power is doubled while the length of the bar (and thus the cost of the semiconductor material) is reduced.

These results are backed by a thorough low power single emitter analysis, in which various Al-contents in the waveguide were assessed. The low current measurement results of SLP single emitters indicate that for a temperature of $T_{HS} \sim 200$ K an Al-content of $x = 4\%$ in the waveguide offers the lowest series resistance without degrading output power. A further reduction of Al-content leads to significant deterioration of the threshold current and slope with little benefit to series resistance. However, tests on FP single emitters to high powers reveal, that the SQW structures with $x = 4\%$ Al-content in the waveguide suffer from strong power saturation, attributed to the low barrier height around the quantum well. Possible options are shown to be the increase of Al-content to $x = 7.5\%$ or the implementation of further QWs, which lead to higher series resistance or higher threshold current, respectively. The first option lowers efficiency at high currents while reaching a high peak efficiency, and the latter limits peak efficiency but sustains high efficiency to high currents.

It was shown that all structures suffer from an excess resistance which does not follow the behavior of the expected bulk resistance in decreasing with temperature, but instead increases. This excess resistance depends heavily on the Al-content in the waveguide and is shown to be caused by the design of the active region. In particular a strong correlation was observed between the barrier height around the active region and the excess resistance, which represents the first report of a resistance contribution from the active region. Changes to the active region such as the introduction of a low p-doping prove to reduce this parasitic resistance and further studies into the origin of this resistance are ongoing.

A thorough analysis of the loss channels to the dissipated power shows, that even the designs with $x = 4\%$ Al-content in the waveguide still possess a series resistance which is too high to achieve efficiencies $\eta_E \geq 65\%$ at the targeted powers. Looking at equation 5.1 thinning of the waveguide is an option to reduce series resistance. Also, increasing the asymmetry of the vertical structure is proven to lead to lower R_S. Together with a low Al-content, high efficiencies at high powers can be expected.

Chapter 6

Low Al-content, thin waveguide designs (iteration 2)

In this chapter the findings from chapters 4 and 5 regarding the efficiency at high powers will be applied to a final iteration 2. As resistance still remains the biggest challenge to high efficiencies even in the low Al-content structures of iteration 1, designs are developed to further reduce this. A lower limit for the Al-content in the waveguide is deduced from iteration 1 and the results from iteration 0 regarding thin, asymmetric waveguides are taken into consideration. Three vertical designs with different waveguide compositions are assessed in a short loop process for their behavior at room temperature before being facet coated and assembled for low temperature testing as well as investigating their electrical properties. The results show a low series resistance and efficiencies > 70% at the highest applied currents.

The performance of the three designs at bar level is discussed in the second part of the chapter. High fill-factor bars and single emitters with two different stripe widths are fabricated using the low Al-content, thin waveguide designs. Measurements confirm the good performance at high currents and at low temperatures in single emitters, but all assembled bars suffer from emitter failures at higher currents. Though low Al-content designs benefit from a much decreased series resistance, power saturation effects at high currents limit the output power. A thorough comparison between single emitters and bars reveals that single emitters are negatively affected by a smaller contact area due the the wafer process specifications, leading to higher threshold current and series resistance. Bars on the other hand suffer from a degraded n-metallization acquired during processing, significantly increasing series resistance and countering the benefit of the low resistance vertical design. Also, output power in the bars is not perfectly distributed across the emitters, leading to slightly lower external differential efficiencies. Efficiency of bars at a power level of 1.5 kW is > 62%, a significant increase compared to SQW structures of iteration 1 and the highest efficiency at this power level yet. An analysis of the remaining loss mechanisms reveals despite the low series resistance, the Ohmic heating is still the major obstacle to higher efficiencies at high powers. This time, however, the increase in operation current due to power roll over is the main challenge and a possible path to suppress roll over is suggested.

6.1 Al-variation in thin waveguide designs

The main drawback of low Al-content SQW structures in iteration 1 was the early, strong onset of power roll over limiting the achievable output power as well as significantly deteriorating the efficiency. The roll over was attributed to the small barrier heights around the active region together with the band bending at higher currents. At the same time, the structure Design2 from iteration 0 with a very thin, asymmetric waveguide showed little roll over. The idea in iteration 2 is to combine the lower limit for the Al-content determined in iteration 1 with the benefits of a thin, asymmetric waveguide observed in iteration 0. However, the possible limitations of this approach should be acknowledged in the beginning, namely the lower performance improvement seen in iteration 0, the higher chances of facet failures due to a lower d/Γ and the higher vertical far field distribution $\Theta_{95\%}^v$. When processed into bars, a thicker p-side protects the active region from mounting stress during p-down assembly. Furthermore, the implementation of possible gratings for wavelength stabilization is much more difficult in a thin, asymmetric waveguide structure.

6.1.1 Vertical design of thin WG structures

A room temperature optimized structure was chosen as the starting point for the design iteration 2. The thickness of the waveguide is about 50% lower than in the structure used in iteration 1. The original structure featured a SQW emitting at 940 nm asymmetrically embedded in an AlGaAs waveguide with 25% Al-content and fast axis beam divergence was $\Theta_{95\%}^v$ = 51.3°. Asymmetry of the waveguide calculated by the ratio of p-WG thickness compared to thickness of the total waveguide (with 50% being the symmetric case) is increased from 40% (iteration 1) to 21%. Also, the doping of the p-WG is strongly increased. To compensate for the small distance between the QW and the p-contact due to the thin p-WG and to counter mounting stress, a thick, highly doped GaAs sub-contact layer was introduced to the p-side. The development of the thin, asymmetric WG structures was performed while the full wafer process from iteration 1 was still ongoing. Due to the temporal overlap, lessons learned from the high power measurements of FP devices from iteration 1 could not be employed in the vertical design development, so that the use of DQWs for improved high power operation and doping of the active zone to reduce excess resistance were not implemented. However, the results from iteration 2 using a SQW in the vertical structure will support a thorough comparison with the various SQW designs from iteration 1.

The original structure was altered to achieve high power, high efficiency operation at 940 nm at temperatures around 200 K. First, the composition of the QW was changed so that emission is around 975 nm at room temperature. Second, the Al-content in the waveguide was decreased to reduce the series resistance while at the same time providing sufficiently high barriers around the QW for low leakage, high power operation. The Al-content was reduced in two steps to 15% and 7.5%. The latter was deduced as the lower limit from iteration 1, in which a SQW embedded in a 4% Al-content WG proved to be too aggressive of an approach resulting in increased power saturation at high currents due to the low barriers around the QW. To show the difference between iteration 1 and iteration 2, the two $x = 15\%$ designs are depicted in figure 6.1.

Figure 6.1 (a) shows the undoped, unbiased band structure of the $x = 15\%$ designs from iteration 1 and iteration 2. While energy levels in the waveguide are identical due to the same Al-contents, the waveguide in iteration 2 is much thinner. Also, the higher asymmetry of the design is visible, e.g. the thinner p-waveguide. While 40% of the waveguide in iteration 1 were p-doped, designs in iteration 2 only feature a p-waveguide of 21% of total waveguide thickness. Invisible to the naked eye in this figure are the changes to the active region, which uses a different layer configuration in iteration 2, namely a thicker quantum well and an alteration of barrier layers around it. While in iteration 1 the Al-content in these barriers was graded at a fixed doping level, the composition is now constant with a non-uniform doping level. The Al-content at the edge of the QW is the same in both iterations, but due to the grading of the barrier layers in iteration 1, a stronger carrier confinement is established.

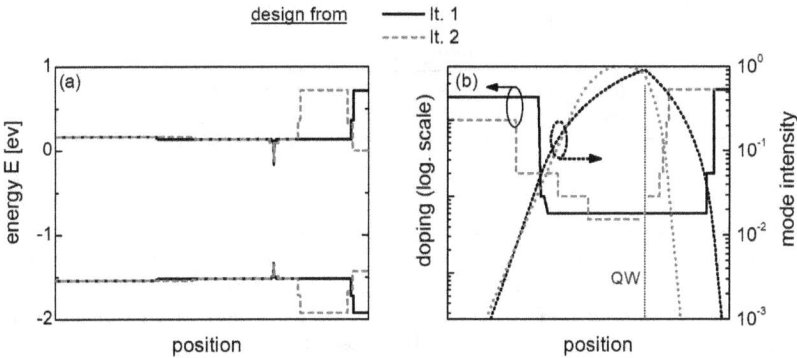

Figure 6.1: Comparison of $x = 15\%$ design from iteration 1 (solid) and iteration 2 (dashed). (a) Conduction and valence band. (b) Doping profile and permitivity (dotted).

In figure 6.1 (b) the difference of doping between iteration 1 and iteration 2 can be seen. The low conductivity of the holes in the p-region is boosted by a higher doping of the p-waveguide. At the same time, doping is slightly lower in the n-waveguide, as conductivity is already high. While a higher doping on the p-side normally strongly increases the internal loss due to the absorption of photons by free carriers, e.g. holes with a high absorption cross section [Pet07], the high asymmetry of the structure decreases the overlap and intensity of the optical mode with the p-region by shifting the mode more to the n-side (see vertical mode intensity in figure 6.1 (b)). Thus, a higher p-doping is feasible in iteration 2.

The simulation tool QIP calculates the internal loss of the structure from the overlap of the optical mode with each layer (cf. equation 5.2). Also, the expected series resistance of a design is determined by adding the contributions of the individual layers following equation 5.1 and using published mobilities from [Sot00]. For the two low-temperature designs with $x = 7.5\%$ and 15% in the waveguide as well as for the original reference structure, the values for Γ, R_S of a 100 μm wide, 4 mm long stripe and α_i are given in table 6.1 compared to the room temperature optimized reference structure with 25%

Al-content in the waveguide and 940 nm operation. Furthermore, the values for the $x = 7.5\%$ and 15% structure from iteration 1 are listed for convenience.

Table 6.1: Calculated values for vertical thin, asymmetric waveguide structures with different Al-content. Series resistance is given for a $L = 4$ mm, 100 µm wide single emitter at room temperature and includes substrate and contact resistance. The reference room temperature optimized structure has $x = 25\%$. The bottom set of values is for structures from iteration 1, taken from table 5.2.

Al [%]	Γ [%]	Γ/d_{QW} [nm^{-1}]	R_S [mΩ]	α_i [cm^{-1}]	λ_{RT} [nm]	ΔE [meV]	$\Delta E/k_B T_{HS}$ (300 K)	$\Delta E/k_B T_{HS}$ (200 K)
7.5	0.542	0.06	10.1	0.50	975	161	6.2	9.3
15	0.551	0.061	11.0	0.42	975	183	7.1	10.6
25	**0.627**	**0.07**	**12.1**	**0.46**	**940**	**265**	**10.2**	**15.4**
7.5	*0.369*	*0.053*	*17.7*	*0.48*	*975*	*171*	*6.6*	*9.9*
15	*0.378*	*0.054*	*19.5*	*0.49*	*975*	*232*	*9*	*13.5*

The increasing Al-content in the waveguide compresses the optical mode, reducing its extent into the cladding layers and the confinement factor goes up. This behavior was already seen in the iteration 1 structures. The expected series resistance of a $w = 100$ µm, $L = 4$ mm single emitter rises with the Al-content. Taking into consideration that the values includes the 5 mΩ contribution of the substrate and contact, the change in R_S between the $x = 7.5\%$ and 25% design corresponds to a 40% increase. Finally, the internal loss is not expected to change significantly with Al-content in waveguide, as the doping levels are the same in all three designs. This assumption neglects free carrier absorption by carriers in the well as suggested by [Kau16]. Compared to iteration 1, the calculated values promise a better performance of the thin, asymmetric structures compared to the Asloc structures with the same Al-contents listed in table 5.2. While α_i is similar, the series resistance is strongly reduced. In the case of the $x = 7.5\%$ structure, R_S (without the substrate resistance) is reduced from 12.7 mΩ to 5.1 mΩ. At the same time, the positive impact of a lower Al-content on the series resistance is less pronounced. Between $x = 7.5\%$ and $x = 15\%$, R_S is reduced by 1.8 mΩ for the Asloc designs from iteration 1 compared to the 0.9 mΩ for the thin, asymmetric waveguide structures. While the thicker QW leads to a higher overlap of the vertical optical mode with the thicker quantum well, e.g. a higher Γ than for the Asloc designs, the relative confinement factor Γ/d_{QW} does not significantly change between the iterations. Threshold current is expected to be increased due to the thicker QW. A higher Γ also further decreases the spot size d/Γ and thus increases the risk of facet failure as stated in [Bot99]. The effective barrier height of the $x = 15\%$ design at 200 K matches that of the $x = 25\%$ design at room temperature. Since the reference structure's QW barriers had been optimized for high power, high efficiency room temperature operation, the $x = 15\%$ thus promises a good performance with little leakage across the barriers at the targeted 200 K.

6.1.2 Internal parameters at room temperature

The two low-temperature vertical designs and the reference structure were grown with MOVPE on 3-inch wafers and short loop processed (SLP) to assess the performance. 100 µm wide emitters on uncoated bars of different lengths were tested under short pulse conditions (1.5 µs, 5 kHz) at $T_{HS} = 293\,\text{K}$ to obtain the internal parameters following the discussion in chapter 2. The wavelength, slope and threshold current were recorded for each length and subsequently, the calculated $1/\eta_d$ and $\ln J_{th}$ are plotted against the cavity length (figure 6.2).

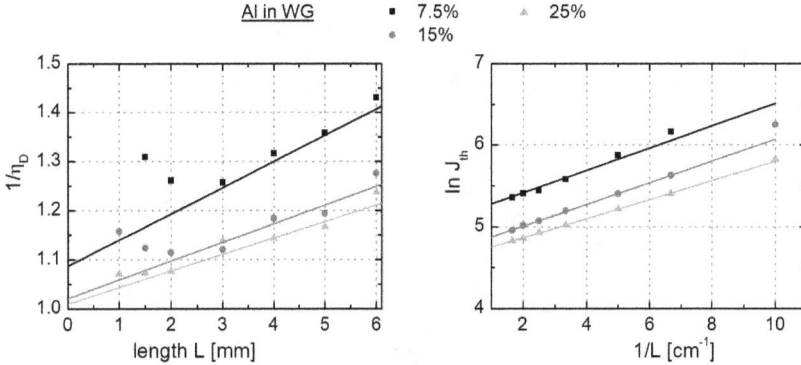

Figure 6.2: Length dependent analysis of thin, asymmetric waveguide structures with different Al-content. (a) Inverse of external differential slope efficiency. (b) Natural logarithm of the threshold current density.

In figure 6.2 (a) the length dependence of the inverse of the external differential slope efficiency allows to gain insights into the impact of Al-content on the internal parameters η_i and α_i using equations 2.9 and 2.15. The intercept of the linear fit with the ordinate increases with decreasing Al-content, i.e. the internal differential efficiency decreases with lower Al-contents. This behavior was also seen in the Asloc structures of iteration 1. However, the length dependent measurements are performed at room temperature, at which point the lower barriers around the active region of low Al-content structures cause significant thermal leakage. This disadvantage was shown to vanish at lower temperatures in chapter 5 as $\Delta E/k_B T_{HS}$ is increased.

From the slope of the fit the internal loss is calculated. While the slope is very similar between the $x = 15\%$ and $x = 25\%$ design, the lowest Al-content structure shows a much stronger dependence of $1/\eta_d$ on length. At the same time, a significant increase in $1/\eta_d$ is observed for cavity lengths $< 3\,\text{mm}$ for the $x = 7.5\%$ design, and the linear fit does not include these measurement points. For the short cavity lengths, α_m gets large (cf. equation 2.5) and a higher gain has to be established by an increased amount of carriers to reach lasing threshold as explained in [Epp13]. The higher amount of carriers in the well and free carriers from leakage (especially prominent in low Al-content waveguide designs, such as the discussed $x = 7.5\%$ design) in turn increase α_i. This explains the

111

difference in measured to calculated α_i in table 6.1. The same effect was observed in low Al-content designs in length dependent measurements in chapter 5.

Figure 6.2 (b) shows the length dependence of threshold current density from which the transparency current density and the modal gain can be deduced. In accordance with the observation made in iteration 1, a lower Al-content in the waveguide leads to an increase in J_T as can be seen from the intercept of the fit with the ordinate. Similar to the behavior of η_i, the transparency current density is expected to become independent of Al-content at lower temperatures with the suppression of leakage currents with an increasing $\Delta E / k_B T_{HS}$. The variation of $\ln(J_{th})$ increases only slightly with lower Al-content, so Γg_0 is increased for low Al-content designs. The calculated confinement factor also decreases with Al-content which explains the change in modal gain between the designs.

The calculated values for η_i, α_i, J_T, Γg_0 and T_0 as well as the measured wavelength of the 2 mm long, 100 µm wide emitter are summarized in table 6.2 for the three Al-contents. For comparison, the values for the $x = 7.5\%$ and 15% designs from iteration 1 are also listed (cf. table 5.3).

Table 6.2: Internal parameters and central wavelength for the three vertical structures with different Al-content in the waveguide for 100 µm wide emitters, measured at $T_{HS} = 293\,K$. For convenience, the values for designs from the previous iteration are given in the bottom section.

Al [%]	η_i [%]	α_i [cm^{-1}]	J_T [A/cm^2]	Γg_0 [cm^{-1}]	T_0 [K]	λ_c [nm]
7.5	92	0.59	160	8.8	58	967
15	98	0.45	109.6	9.1	115	955
25	99	0.40	98.9	10.3	171	937
7.5	*90*	*0.41*	*93.8*	*6.6*	*58*	*974*
15	*98*	*0.37*	*75*	*7.7*	*110*	*973*

The values in table 6.2 confirm the conclusions drawn from figure 6.2. With decreasing Al-content, differential internal efficiency decreases and transparency current density increases, both of which are attributed to increased thermal leakage, as seen in iteration 1. The difference in modal gain is due to the change in confinement factor, seen in table 6.1. Finally, there is an increase in internal loss for the $x = 7.5\%$ structure, while the other Al-content designs show the expected α_i. A possible explanation for the higher internal loss is the additional absorption of free carriers in the quantum well as suggested by [Kau16]

$$\alpha_{QW} = \Gamma (\sigma_n + \sigma_p) N_{th} \tag{6.1}$$

where $\sigma_{n,p}$ are the absorption cross sections for electrons and holes, respectively, and N_{th} is the threshold carrier density in the well. The $x = 7.5\%$ design suffers from a strongly increased number of carriers in the quantum well at threshold, which can already be seen in J_T. The corresponding free carrier absorption in the quantum well leads to an additional contribution to the internal loss, which is expected to decrease at lower temperatures, when leakage currents are suppressed and threshold current is strongly reduced. This effect of QW absorption and absorption due to leakage was not as prominent in the iteration 1, as the barrier height was smaller and the barriers were not graded, which manifests in J_T varying less with Al-content (though it could be observed for the Asloc4SQW design with the low barriers).

Comparing the internal parameters gained from length dependent measurements at room temperature of iteration 1 and iteration 2 of the same Al-contents, the difference in barrier height even for the same Al-content as well as the different barrier layout has to be considered (cf. table 6.1). The differential internal efficiency is the same for the $x = 15\%$ design and differs only slightly for the structure with 7.5% Al-content in the waveguide. This is contrary to expectation, as the highly doped, thick sub-contact layer of designs in iteration 2 was presumed to increase lateral current spreading and thus decrease η_i. Internal loss is similar as well for the higher Al-content design ($0.37\,\mathrm{cm}^{-1}$ iteration 1 compared to $0.45\,\mathrm{cm}^{-1}$ from iteration 2) but differs significantly for the $x = 7.5\%$ design (with $0.41\,\mathrm{cm}^{-1}$ in iteration 1) as explained above. The designs in iteration 2 show a much increased transparency current density, with values around 50% higher than in the corresponding structures in iteration 1. This is attributed to the lateral current spreading in the p-sub-contact layer used in iteration 2. Decreased leakage at lower temperatures and a lateral confinement for the current in p-layers are expected to significantly decrease J_T. The relative confinement factor Γ/d_{QW} was increased by 14% in iteration 2 compared to designs from iteration 1 and the modal gain increased in turn by 18% and 33% for the $x = 15\%$ and 7.5% structures, respectively.

6.1.3 Electrical properties and behavior at low temperatures

The length dependent short pulse measurements could only be performed at room temperature and did not allow precise insights into the electrical properties of the designs, e.g. the series resistance. In order to analyze the low temperature performance and determine the benefit of the thin, asymmetric waveguide on R_S of the three designs with different Al-content in the waveguide, 4 mm long bars from the SLP are facet coated with front and rear facet reflectivities of 2% and 95%, respectively. Facet passivation is omitted, as the devices are to be analyzed in the low current regime. Furthermore, the SLP is generally not designed for high power operation, as explained in the appendix. Single emitter with 100 μm wide stripes are cleaved from the coated bars and mounted p-side up onto CuW heatspreaders using AuSn solder. The Al p-contact is connected with Al bond wires.

Diodes with the different designs are subsequently tested under QCW conditions (1.2 ms, 10 Hz) at $T_{HS} = 288\,\mathrm{K}$ up to $I = 8\,\mathrm{A}$ and at 203 K up to 12 A. The voltage and the optical power are recorded and are displayed together with the calculated efficiency in figure 6.3.

From the room temperature measurement shown in figure 6.3 (a) the impact of Al-content in the waveguide on the electro-optical performance can clearly be seen. Of the three structures, the $x = 7.5\%$ design has the highest threshold current and lowest slope in accordance with the short pulse measurements and the power current characteristics starts rolling over at a low current of $\sim 3\,\mathrm{A}$. Between the $x = 15\%$ and 25% designs, no difference in threshold current can be seen. The observed higher slope of the $x = 25\%$ structure is mostly due to its higher photon energy as it lases at a central wavelength of $\lambda_c = 933\,\mathrm{nm}$ shortly above threshold, compared to $\lambda_c = 955\,\mathrm{nm}$ and $967\,\mathrm{nm}$ for the $x = 15\%$ and 7.5% designs, respectively. By linearly fitting the power-current characteristics shortly above threshold in the interval $[I_{th} + 0.5\,\mathrm{A}, I_{th} + 2.5\,\mathrm{A}]$, the slope is derived and using the measured λ_c, the differential external efficiencies are calculated. Here, the difference between the $x = 25\%$ and 15% designs is little with $\eta_d = 82\%$ and 80%, respectively, while the lowest Al-content structure suffers from a deteriorated differential external efficiency of 68%. Both structures with the higher Al-contents experience no visible roll over up to the maximum current.

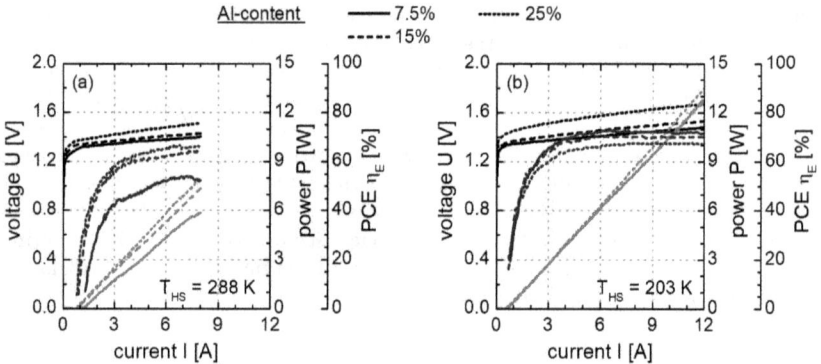

Figure 6.3: Power, voltage and efficiency as a function of current of short loop processed, asymmetric thin waveguide designs with varying Al-content in the waveguide at (a) $T_{HS} = 288\,\mathrm{K}$ and (b) $203\,\mathrm{K}$.

The different lasing wavelengths of the three structures cause the voltage curves to be shifted compared to each other, e.g. have a different intersection U_0 of the linear fit to the voltage current curve with the ordinate. For example, the $x = 7.5\%$ design with the longest wavelength, e.g. the smallest band gap, has the lowest voltage at any current. Vice versa the $x = 25\%$ design, emitting at $933\,\mathrm{nm}$ has the highest U_0. Apart from this parallel shift, the series resistance increases with increasing Al-content in the waveguide, as can be seen by the slope of the UI curve. A linear fit in the interval confirms this observation with $R_S = 22\,\mathrm{m\Omega}$, $14.2\,\mathrm{m\Omega}$ and $14.5\,\mathrm{m\Omega}$ for the $x = 25\%$, 15% and 7.5% structures, respectively. The choice of the fitting interval explains why the series resistance of the 15% Al-content structure is unexpectedly lower than for the 7.5% design. The voltage current characteristics for the diodes of the two designs run parallel at small currents but diverge at higher currents. Thus, the fit at small currents leads to similar R_S values, whereas at higher currents, the low Al-content design shows the expected lower series resistance.

The characteristic parameters of the different vertical designs determined at $T_{HS} = 288\,K$ and at 203 K are listed in table 6.3. For comparison, the corresponding values from iteration 1 are also shown (though the lower temperature was 218 K due to the earlier stage of the measurement setup).

Table 6.3: Measured characteristic parameters for mounted short loop processed BAL from iteration 2 with $L = 4\,mm$, $w = 100\,\mu m$ at two temperatures. For convenience, the corresponding values for designs from the previous iteration are given in the bottom section.

Al [%]	η_d [%] (288 K)	η_d [%] (203 K)	I_{th} [mA] (288 K)	I_{th} [mA] (203 K)	R_S [mΩ] (288 K)	R_S [mΩ] (203 K)	λ_c [nm] (288 K)	λ_c [nm] (203 K)
7.5	68	86.7	1060	650	14.5	14	967	940
15	80	83.1	800	570	14.2	18.5	955	929
25	82	83.4	740	600	22	28.6	933	907
	(288 K)	(218 K)	(288 K)	(218 K)	(288 K)	(218 K)	(288 K)	(218 K)
7.5	*77*	*88.8*	*750*	*440*	*16.9*	*21.1*	*983*	*958*
15	*84*	*89.4*	*580*	*430*	*29.8*	*34.5*	*977*	*954*

Comparing the $x = 7.5\%$ and 15% Asloc designs from iteration 1 to those of iteration 2, series resistance at room temperature for the thin, asymmetric waveguide structures is decreased from 16.9 mΩ and 29.8 mΩ to 14.5 mΩ and 14.2 mΩ, respectively. This is due to the thinner waveguide, especially the thinner p-side. At the same time, external differential efficiency is reduced from 77% and 84% for the $x = 7.5\%$ and 15% structures from iteration 1 to $\eta_d = 68\%$ and 80.3% in iteration 2. The internal parameters gained from the short pulse measurements show very similar values for η_i and α_i in iteration 1 and iteration 2 lead to expected similar values of η_d, though. This suggests an overestimation of one of the parameters in iteration 2. As mentioned above, η_i in designs from iteration 2 is expected to be lower due to strong lateral current spreading in the highly doped, thick sub-contact GaAs p-layer. The absolute values for η_i are thus overestimated in the short pulse measurements, probably due to inaccurate power calibration.

Threshold current is higher for the thin waveguide designs with $I_{th} = 1060\,mA$ and 800 mA for the $x = 7.5\%$ and 15% structures compared to 750 mA and 580 mA in iteration 1. This was expected from equation 2.8 with the values gained from the length dependent analysis, in which a higher transparency current density for iteration 2 designs was found. The difference in I_{th} is less pronounced than the difference in J_T due to the higher modal gain in structures in iteration 2.

The performance at the lower temperature of $T_{HS} = 203\,K$ of the three designs is displayed in figure 6.3 (b). No difference in threshold current is visible between the structures. Also, slope of the lower Al-content designs is the same, consistent with the suppression of leakage currents for the $x = 7.5\%$ design. As wavelength between the structures differs, the external differential efficiency is calculated from the linear fit to the power-current characteristics in the interval $[I_{th} + 0.5\,A, I_{th} + 2.5\,A]$ and the central wavelength. While the slope of the highest Al-content design is the steepest, the lowest Al-content structure shows the best differential external efficiency after accounting for the different wavelengths with η_d = 86.7%, followed by the x = 25% design with 83.4%. The design with 15% Al in the waveguide has the lowest η_d with 83.1%.

All designs show no roll over in the power-current characteristic up to the maximum current of I = 12 A. Thus, the lower Al-contents cause no deterioration to the output power at this temperature. At the same time, the benefit to the bias voltage is clearly visible. Omitting the parallel shift in voltage due to the difference in lasing wavelength, the low Al-content structures show a lower slope to the UI curve. A linear fit confirms a reduced series resistance with R_S = 14 mΩ, 18.5 mΩ and 28.6 mΩ for the x = 7.5%, 15% and 25% designs, respectively. Together, the high output powers and the low R_S lead to a much increased efficiency across all structures. With $\eta_{E,max}$ = 73% the x = 7.5% design has the highest peak efficiency, followed by 71% and 67% by the x = 15% and 25% structures, respectively. At higher currents, the higher efficiency of the lowest Al-content structure is expected to deteriorate less than for the other structures due to the lower series resistance.

The comparison of designs from iteration 1 and iteration 2 at low temperatures reveals once again the benefit of the thin, asymmetric waveguide structures. While the SLP diodes in iteration 2 were measured at T_{HS} = 203 K, the lowest assessed temperature for the equivalent lasers from iteration 1 was 218 K. The difference in T_{HS} of 15 K will lead to a slightly overstated I_{th} and an undervalued η_d and R_S in iteration 1. Nonetheless, differences between the iterations should still be correctly identifiable and are listed in table 6.3.

Series resistance is greatly reduced from 21.1 mΩ and 34.5 mΩ for the x = 7.5% and 15% design from iteration 1 to 14 mΩ and 18.5 mΩ in iteration 2, respectively. Taking into account the 5 mΩ contribution of the substrate universal to all diodes, the change in design results in a 44% and 54% reduction in series resistance for the 7.5% and 15% Al-content structure, respectively. The external differential efficiency is lower in iteration 2 with 86.7% and 83.1% in the x = 7.5% and 15% case, compared to 88.8% and 89.4% in iteration 1. As discussed for the room temperature measurements, the increase of lateral current spreading in the sub-contact p-layer is assumed to cause a decrease in η_i in designs from iteration 2, leading to a reduced η_d. Finally, the lower threshold current of iteration 1 diodes already observed at room temperature is still present at the low T_{HS} with I_{th} = 440 mA and 430 mA for the x = 7.5% and 15% designs in iteration 1 compared to 650 mA and 570 mA for the respective designs in iteration 2. The introduction of lateral current confinement in the highly doped p-layers in a full wafer process is expected to significantly reduce the lateral current spreading, which is responsible for the high values of I_{th}.

Up to the assessed currents, the reduction of Al-content in the thin waveguide leads to the highest efficiencies yet. However, the behavior of these structures up to the high operation currents is still unknown and will have to be analyzed with fully processed, high power single emitters and bars.

6.2 High power operation of bars and single emitters

In the previous section, the benefit of thin, asymmetric waveguides with low Al-contents for the performance at low temperatures was shown. In order to assess their behavior at the highest currents and powers, the three discussed designs are processed into high power single emitters and bars. A thorough comparison between the bars and single emitters over a wide range of temperatures is performed to track the impact of the low Al-contents and check for the quality of the bars. At the highest currents, roll over in the lower Al-contents is observed and all bars suffer from failing emitters. Based on the single emitter results an estimation of the most important loss mechanisms is performed and options for future improvements in the structure are discussed.

6.2.1 Fabrication of high power bars and single emitters

The three vertical designs featuring a thin, asymmetric waveguide with Al-contents of 7.5%, 15% and 25% are again grown on 3 inch GaAs substrate wafers using MOVPE. For the two low Al-content structures, the composition of the quantum well is adjusted for an emission at $\lambda = 940$ nm at the operation point and temperature of ~ 200 K. As the high Al-content of the $x = 25\%$ baseline structure is not optimized for low temperatures, the quantum well remains unchanged to achieve room temperature high power bars emitting at 940 nm. The wafers are subsequently laterally structured in a full laser process into single emitters and bars with the same layout used in iteration 1:

- Bars with FF = 69 %: 37 stripes with $w = 186$ μm and $L = 4$ mm, pitch 250 μm.

- Single emitters with $L = 2$ mm of widths 100 μm and 186 μm, pitch 550 μm.

The deteriorating effect of lateral current spreading in the p-sub-contact layer was corrected in the FP by using a highly energetic He implantation next to the stripe (150 keV compared to the 50 keV in the SLP), laterally limiting the current path through the p-layers. Plasma-etched index guides in distance of $5\,\mu$ of the stripes extend down to the p-waveguide and further limit the lateral current spreading. Thus, higher internal and thus external differential efficiencies are expected from FP diodes than from SLP diodes.

During the wafer process, problems occurred during the n-side metallization. Not all wafers were metallized in the same run, but in three different batches. Wafers of the last metallization batch performed so poorly in a solder test, the n-metal had to be removed and redone. The wafers with the $x = 7.5\%$ design were in one (referred to as Run1), the structures with 15% and 25% Al-content in another run (referred to as Run2). Only after being subjected to high cleaning temperatures in the facet passivation process signs of a structural change of the n-side metal could be seen under the microscope, most clearly in a dark field image (figure 6.4). In all passivated bars, particles appear in the n-side gold, suggesting a diffusion from lower layers into the contact due to the exposure to the facet cleaning temperatures of 350 °C during passivation. The amount of particles differs between bars from metallization Run1 and Run2, as can be seen in the dark field image with a magnification of 500. In subsequent studies at the institute not connected to this work, the thickness of the n-side diffusion barrier and the cleaning temperatures were altered to prevent particles from lower layers to surface.

117

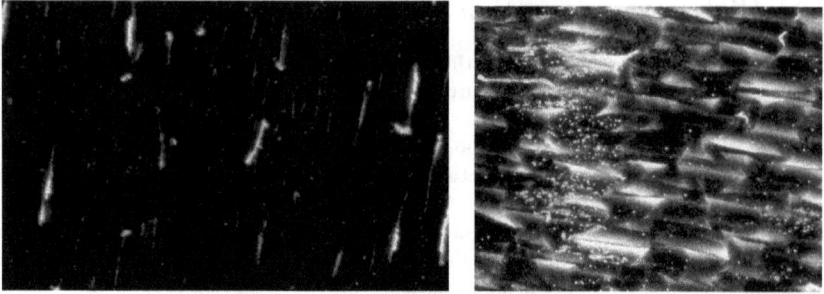

Figure 6.4: Dark-field image with 500x zoom of n-side metal after being subjected to temperatures of 350 °C during the facet passivation process. Left: wafer with $x = 7.5\%$ design from metallization Run1. Right: wafer with $x = 25\%$ design from Run2.

The degraded n-contact will likely lead to an extra contribution to series resistance, which will be wafer dependent. The superposition of the varying n-contact resistance and the series resistance of the semiconductor material will make the interpretation of the effect of Al-content in the waveguide and the benefit of thin, asymmetric waveguides more challenging. Due to the higher pitch of the bars with the single emitter layout, the deteriorated n-contact will further impact the series resistance less than in the high fill factor bars.

Bars with the single emitter layout and $L = 4\,\text{mm}$ are cleaved from the wafer. Though the cavity length of these bars is 4 mm, the contact length is in fact shorter. For all diodes on the wafer, bars and single emitters, the contact is pulled back from each facet by 100 µm by implanting the stripe and hindering current flow to catastrophic optical mirror damage [Die00]. While the bars' $L = 4\,\text{mm}$ long cavity is continuously covered by the p-contact metal, the single emitter is comprised of two $L = 2\,\text{mm}$ long cavities. Thus, the bars' contact length is reduced by 200 µm from the two facets, e.g. 5% of the total cavity length, whereas the single emitter have a 10% smaller contact length, as a 200 µm long middle part is also not accessible for the current. This feature of the uncontacted middle section has to be considered when comparing bars to single emitters.

The facets of bars of all vertical and lateral designs are first passivated with ZnSe for high power operation and subsequently coated with front and rear reflectivities of $R_\text{f} = 2\%$ and $R_\text{r} = 98\%$, respectively. The lessons from iterations 0 and 1 regarding the low package resistance assembly of bars are implemented and bars of all vertical designs are fabricated. First, the bar is sandwiched between two CuW headspreaders using AuSn solder. In a next step, the sandwich is soldered p-down onto a CCP mount using In and a massive n-contact is attached, providing a low resistance electrical connection.

Single emitters with $w = 100\,\text{µm}$ and 186 µm are soldered p-side down onto CuW heatspreaders and contacted with bond wires on the n-side.

6.2.2 Comparison of short loop and fully processed diodes

First, the FP SE diodes with $w = 100\,\mu\text{m}$ and $L = 4\,\text{mm}$ are compared to the SLP results. The direct comparison will help to verify the following two expectations, derived from the process properties of the FP:

- Higher η_d in FP diodes due to decreased lateral current spreading.

- Higher R_S in FP diodes due to degraded n-metallization, dependent on wafer.

Since all these parameters are gained from linear fits shortly above threshold, FP single emitters of the different vertical designs are tested at temperatures between $T_\text{HS} = 298\,\text{K}$ and $208\,\text{K}$ (with $288\,\text{K}$ the highest temperature for the $x = 7.5\%$ design) in QCW mode ($1.2\,\text{ms}$, $10\,\text{Hz}$). The values for η_d are gained from the linear fit to the power-current curve in the interval $[I_\text{th} + 0.5\,\text{A}, I_\text{th} + 2.5\,\text{A}]$ with the knowledge of the wavelength shortly above threshold. R_S is determined by linearly fitting the voltage-current curve in the same interval. The two characteristic parameters are plotted for each T_HS in figure 6.5 next to the values from the SLP diodes (cf. table 6.3) at the two temperatures $288\,\text{K}$ and $203\,\text{K}$.

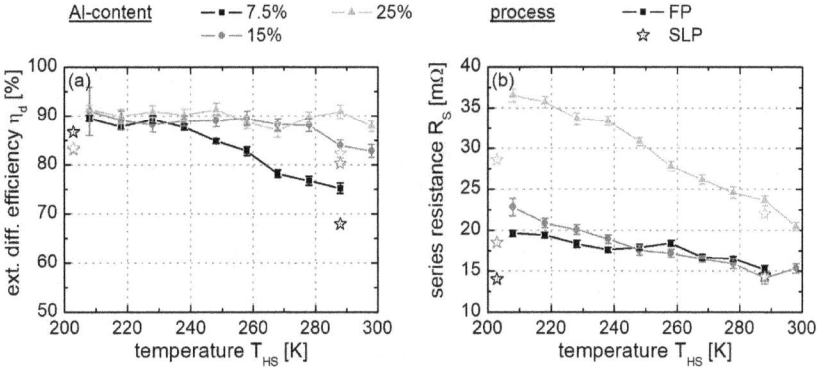

Figure 6.5: Characteristic parameters for FP (solid) and SLP (open stars) diodes with different Al-content in the waveguide plotted against T_HS. (a) external differential efficiency and (b) series resistance.

From figure 6.5 (a), it can clearly be seen that the external differential efficiency is significantly increase in FP diodes compared to the SLP chips. The assumption of an increased lateral current spreading in the thick p-sub-contact layer, which decreases η_i is thus confirmed and current spreading is successfully reduced in FP diodes by a deeper implantation as well as the lateral index trenches next to the stripe.
The temperature behavior of the series resistance is plotted in figure 6.5 (b). While R_S values at room temperature are similar between SLP and FP diodes, they are much higher for the FP chips at lower temperatures. The strongest difference in R_S is seen for

the $x = 25\%$ design, which showed the worst n-metallization (cf. figure 6.4). Thus, the deteriorating impact of the n-metallization from the FP is shown.

Further, the characteristic parameters for the fully processed diodes from iteration 1 and iteration 2 using SQWs with $w = 100\,\mu m$ and $L = 4\,mm$ are compared at the lowest measured temperature in table 6.4.

Table 6.4: Characteristic parameters of $w = 100\,\mu m$ and $L = 4\,mm$ diodes using SQW and different Al-contents in the waveguide at 208 K

Al [%]	η_d [%]	I_{th} [mA]	R_S [mΩ]
7.5	90	378	19.6
15	91	385	22.9
25	91	386	36.5
4	*88*	*515*	*15.8*
7.5	*89*	*547*	*26.2*

The values for designs from iteration 2 in table 6.4 confirm the assumption that threshold current and external differential efficiency are independent of Al-content in the waveguide, i.e. barrier height. Comparing designs from iteration 1 and iteration 2 shows the benefit of the thin, asymmetric waveguide design. While η_d is similar between the designs of the two iterations, series resistance is significantly reduced in iteration 2 for the comparable $x = 7.5\%$ design. Furthermore, the higher modal gain seen in designs from iteration 2 leads to a lower I_{th} when leakage as well as lateral current spreading are minimized by the low temperature and the lateral current confinement, respectively.

6.2.3 Temperature dependent performance of bars and single emitters

The fabricated bars are first checked for their near field uniformity to assess if emitters are working properly by imaging the single emitters with a telescopic lens setup into an integrating sphere and determining the power distribution across the bar at room temperature operated in QCW mode (1.2 ms, 10 Hz) at a fixed, low current of $I = 0.4\,kA$. From each design, the mounted bar with the most evenly distributed power is chosen and subsequently tested over a wide range of temperatures ($T_{HS} = 203\ldots288\,K$) to track the benefit of the thin, asymmetric waveguide and the impact of the Al-content on bar level. These tests are performed under QCW condition (1.2 ms, 10 Hz) only up to low currents of $I = 0.4\,kA$ as not to damage any of the emitters on the bar. Ideally, a bar behaves like the sum of the single emitters it is comprised of. To get an upper limit for the bar performance at the different temperatures, single emitters of the three vertical designs are tested at the same operating conditions as the bars up to a maximum current of $I = 20\,A$ in a temperature range $T_{HS} = 208\ldots288\,K$ with only the lowest operation temperature differing slightly from the bars'. Good comparability is achieved between the devices, as single emitters are used with the same lateral layout present on the bar, e.g. 186 µm wide and $L = 4\,mm$ long. The single emitter results are subsequently scaled by the number of emitters on the bar.

By linearly fitting the bar and single emitter electro-optical results in the interval $[I_{th} + 20\,A, I_{th} + 95\,A]$ (over 15 measurement points) and $[I_{th} + 0.5\,A, I_{th} + 2.5\,A]$, respectively, the characteristic parameters of the designs are extracted for the two assembly configurations. The temperature dependence of the characteristic parameters of the bars and single emitters is compared in figure 6.6.

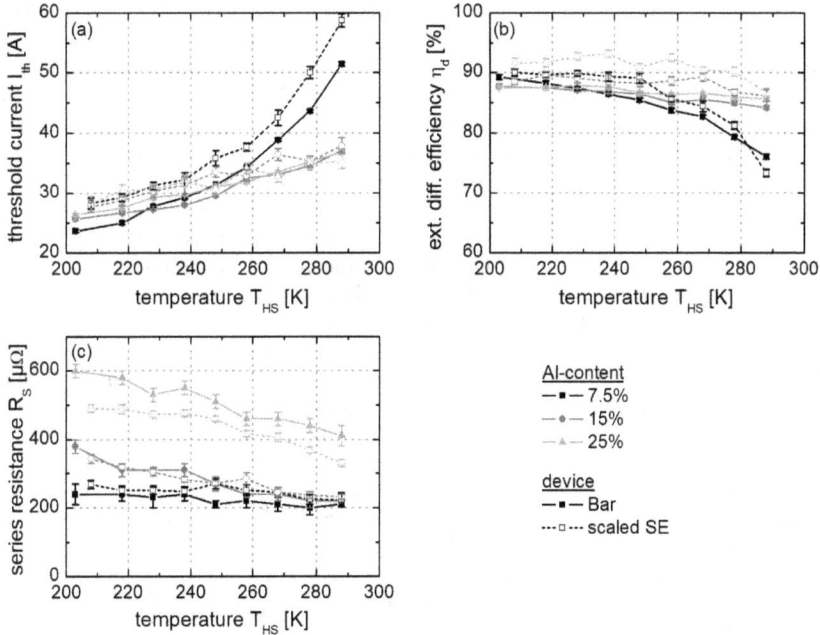

Figure 6.6: Comparison of characteristic parameters from bars and scaled single emitters with thin, asymmetric waveguides and different Al-contents over a wide temperature range. (a) Threshold current; (b) external differential efficiency and (c) series resistance.

In figure 6.6 (a), the threshold currents of the bars and scaled single emitters using the three different vertical designs are plotted against the heatsink temperature. For all designs and temperatures, the threshold current of the single emitters is always higher than the threshold current of the bar (except for four higher temperatures for the $x = 25\%$ design). The difference is most prominent in the $x = 7.5\%$ structure. Here, the bars have a $\Delta I_{th} = 7.3\,A$ lower threshold current at $T_{HS} = 288\,K$ decreasing to $\Delta I_{th} = 4.6\,A$ at $T_{HS} = 218\,K$. For the $x = 15\%$ design, the threshold current of the bars is $1\,A$ lower at room temperature than the single emitter's and $1.3\,A$ lower at $T_{HS} = 218\,K$. While I_{th} is very similar between bar and single emitter at warmer temperatures in the case of the design using 25% Al-content waveguides with the single emitter even having a slightly lower threshold current, the bar shows a visibly lower I_{th} at temperatures below $T_{HS} = 238\,K$. The discrepancy in threshold current is caused by the mentioned difference in contact length of the bars and single emitters of same cavity length. In the unpumped middle part of the single emitters the transparency of the semiconductor has to be established by optical absorption, thus increasing the optical loss. This saturable absorber in the middle of the cavity increases the threshold current of the device. Above threshold, the optical loss is pinned and is the same in the bar and the single emitter. The external differential efficiency of the bars and single emitters is compared in figure 6.6 (b) for different temperatures. Across all diodes, η_d is increased at low temperature compared to room temperature. As expected, the low Al-content design shows the strongest rise in η_d, going from 73.3% and 76% at T_{HS} = 288 K to 90% at 208 K and 89.2% at 203 K for the single emitter and bar, respectively. The bar with the $x = 15\%$ structure shows an increase from $\eta_d = 84.2\%$ to 87.7% from highest to lowest temperature and thus surpasses the bar with the highest Al-content, which experiences an increase from 85.6% to 87.6% in the same temperature interval. This is in contrast to the results of the single emitter measurements. Here, the $x = 25\%$ design has the highest external differential efficiency at all temperatures. However, below T_{HS} = 258 K, η_d does not rise any more, but in fact decreases slightly to 91.6% at the lowest temperature. This behavior of deteriorating external differential efficiency at the lowest temperatures had previously also been observed in chapter 4 in the single emitters of Design2, which featured a similar Al-content and asymmetry of the waveguide. The single emitter with the $x = 15\%$ design shows an external differential efficiency of 86.2% at room temperature with the maximum $\eta_d = 89.6\%$ at 218 K, after which point it drops back to 88.5% at 203 K. Between the two 975 nm structures (with $x = 7.5\%$ and 15%), no significant difference in η_d can be made out at lower temperatures. This is in accordance with the results in chapter 5, in which external differential efficiency becomes independent of Al-content at low temperatures. Comparing η_d of the single emitters and bars, the single emitters generally show higher values than the bars. This is in line with the assumption, that a bar ideally behaves like the sum of its emitters. If any of the single emitters the bar is comprised of show a poorer than ideal performance, the results from single emitter and bar will differ. In table 6.5 the average power per emitter on the bar at a current of 400 A is listed as measured at room temperature with the telescopic lens setup and compared to the output power of the single emitter at the equivalent current level.

Table 6.5: Power of single emitter and average power per emitter on bars with different Al-contents in the waveguide, measured at T_{HS} = 288 K at a bar current of 400 A.

Al [%]	P_{SE} [W]	P_{bar} [W]	Δ_{rel} [%]
7.5	8.55	8.4	1.8
15	10.8	10.43	3.4
25	11.38	10.95	3.8

The relative difference between the power of the single emitter and the average power of the emitters on the bar listed in table 6.5 corresponds well to the difference in external differential efficiency seen in figure 6.6 (b) between bars and single emitters. Thus, the reduction of η_d witnessed in bars is likely to stem from non ideal emitters.

Figure 6.6 (c) shows the series resistance for the bars and scaled single emitters for the different designs at different temperatures. Generally the series resistance rises with decreasing temperature for all devices. In agreement with the findings in chapters 4 and 5, the absolute resistance as well as its increase with temperature depend on the Al-content in the waveguide and the $x = 7.5\%$ design shows the lowest values. R_S for the bar and single emitter with this low Al-content design rises from $0.21\,\text{m}\Omega$ and $0.22\,\text{m}\Omega$ at room temperature to $0.24\,\text{m}\Omega$ at $203\,\text{K}$ and $0.27\,\text{m}\Omega$ at $208\,\text{K}$, respectively. The higher series resistance found in the single emitter stems from the shorter contact length. This difference can only be seen in the diodes with a good n-metallization, e.g. with the $x = 7.5\%$ design, as increased resistance due to the metallization is superimposed. While in the $x = 15\%$ case, R_S is roughly the same between single emitter and bar, rising from $0.22\,\text{m}\Omega$ at room temperature to $0.38\,\text{m}\Omega$ at $203\,\text{K}$, the bar with the $x = 25\%$ design shows a significantly higher series resistance, increasing from $0.41\,\text{m}\Omega$ at room temperature to $0.6\,\text{m}\Omega$ at $203\,\text{K}$, compared to $0.33\,\text{m}\Omega$ at $288\,\text{K}$ and $0.49\,\text{m}\Omega$ at $208\,\text{K}$ for the equivalent scaled single emitter. The difference in series resistance between bars and single emitter stems from the degraded n-side contact. While the emitters on the bar have a pitch of $250\,\mu\text{m}$, the mounted single emitter chips are $550\,\mu\text{m}$ wide (both with a p-contact width of $186\,\mu\text{m}$). Thus, the faulty n-contact area is much wider for the single emitter, reducing resistance.

6.2.4 High power performance of bars with thin, asymmetric waveguide designs

After the comparison of bars and single emitters at low powers and different temperatures, the devices are driven up to high currents of $I = 2\,\text{kA}$ and $I = 50\,\text{A}$ at a low temperature. QCW test conditions were chosen as $\tau = 200\,\mu\text{s}$, $f = 10\,\text{Hz}$ for the bars as devices were tested at the lower temperature limit of the measurement station and higher amounts of waste heat due to longer duty cycles could no longer be removed. $125\,\mu\text{s}$, $16\,\text{Hz}$ pulses were chosen for the single emitters due to the specifications of the current supply, to match the low duty cycle of 0.2% of the bar measurements. The maximum current applied to the diodes was in both cases limited by the maximum output current of the respective current supply. The heatsink temperature for these high current tests was set to match the lowest T_{HS} of the respective temperature series, $203\,\text{K}$ for the bars and $208\,\text{K}$ for the single emitters. Thus, a small discrepancy of $5\,\text{K}$ exists between bar and single emitter measurement. However, this small difference is not expected significantly impact the comparability between bar and single emitter. Finally, after the high current test, another histogram of the bars is recorded at room temperature at the same conditions as the power distribution analysis to examine for possible emitter failures.

The performance of the bar and scaled single emitter using the $x = 7.5\%$, thin, asymmetric waveguide design is compared in figure 6.7 (a). The bar reaches a peak efficiency of 77% and sustains a $\eta_E > 62\%$ up to a power of $1.5\,\text{kW}$, the highest achieved bar peak efficiency and at this power level yet. However, the early onset of roll over above $I = 0.8\,\text{kA}$ strongly limits output power and decreases efficiency. The failure of at least one emitter at $I = 1.72\,\text{kA}$, visible by the kink in the power current characteristic further decreases output power. Thus, the bar only reaches a maximum output power of $1.88\,\text{kW}$ where efficiency is 53%. The scaled single emitter behaves very similarly, as was predicted from the analysis of characteristic parameters of the bar and single emitter at different temperatures. Analyzing the shift in wavelength with increasing waste heat, the thermal resistance for the bar and single emitter is found to be 0.006 K/W and 0.0062 K/W, respectively. A visible difference is the higher threshold current of the single emitter which causes a parallel shift of the power-current characteristic (slope is the same as predicted by a similar η_d) and in turn a reduced efficiency. Simultaneously, the higher series resistance due to the reduced contact length is visible in the UI curve. Peak efficiency is lower with 74% and also at a scaled power of 1.5 kW with 59.2%. A maximum power of 1.8 kW is reached, limited by the available current of the source. While originally output power and efficiency are lower at every current for the scaled single emitter than for the bar, the mentioned emitter failure in the bar leads to an identical output power and efficiency from this current on. The non-thermal roll over of the two devices is similar and compares to the high power behavior of the Asloc4SQW design seen in iteration 1. Following the argumentation in the previous chapter, the low barriers around the QW due to the low Al-content of 7.5% lead to increased leakage at high currents, negatively impacting the output power. As effective barrier height at 200 K is lower for the $x = 7.5\%$ design from iteration 2 than from iteration 1, the leakage and thus the power saturation at high currents is increased.

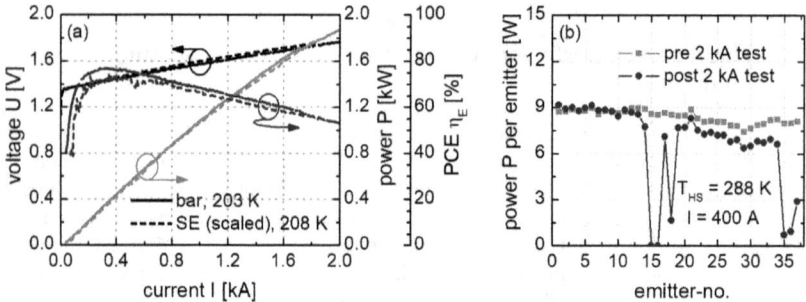

Figure 6.7: (a) Electro-optical characteristics of a bar at $T_{HS} = 203$ K and a scaled single emitter at 208 K both with 7.5% Al-content in the thin, asymmetric waveguide. (b) Distribution of power prior and after the test up to $I = 2$ kA.

In figure 6.7 (b), the power distribution across the bar is compared at room temperature at a low current before and after the test up to the highest currents. While there is a slight deterioration of output power observed from one side of the bar to the other, all emitters are working before the high power measurement. An average power of 8.4 W per emitter is recorded with a standard deviation of the mean $\sigma_P = 0.42$ W. The single emitter power at an equivalent current is 8.6 W. After the high current test, however, multiple emitters are damaged. A total of six emitters is completely or partly impaired while the reduction of output power of one side of the bar is increased. The high number of failed emitters compared to iteration 1 is potentially caused by the lower spot size due to the higher confinement factor and thinner waveguide, increasing the facet load and the sensitivity to contaminations or defects on the facet.

The electro-optical characteristics of a bar and a scaled single emitter with the $x = 15\%$ Al-content design are displayed in figure 6.8 (a). Peak efficiency in the bar is 75%, dropping to 62% at a power of 1.5 kW. At the maximum current, a power of 1.95 kW is reached with $\eta_E = 53\%$. Output power at high currents is reduced by the onset of power roll over, starting at $I \sim 1$ kA. Also at that current, a kink in the power current characteristic indicates an emitter failure with further drops in power observed at 1.49 kA and 1.83 kA. The bar behaves very similarly to the scaled single emitter results having the same slope up to the point of the first emitter failure. This in in line with the comparison of characteristic parameters of single emitter and bar, in which a the devices showed the same R_S and η_d and single emitters had only a slightly increased threshold current. The latter feature is barely visible in the power current characteristics, but translates into a marginally lower peak efficiency of 74.5% for the single emitter. Thermal resistance is alike with $R_{th} = 0.0089$ K/W for the bar and 0.0085 K/W for the single emitter. From the second emitter failure on the bar onwards, the two power current curves start to diverge, as the remaining emitters on the bar show a stronger roll over than the scaled single emitter. The scaled single emitter predicts an output power of 1.95 kW with $\eta_E = 60\%$ in case of a bar with all emitters working.

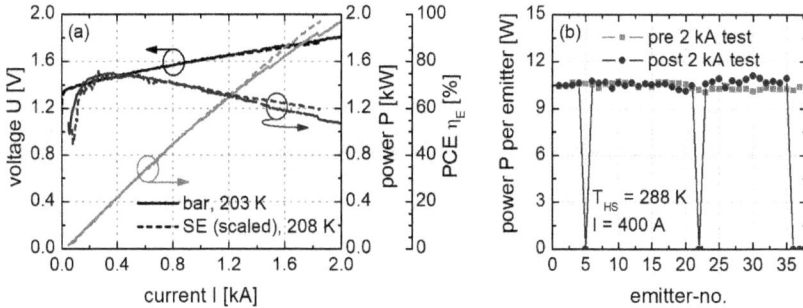

Figure 6.8: (a) Electro-optical characteristics of a bar at $T_{HS} = 203$ K and a scaled single emitter at 208 K both with 15% Al-content in the thin, asymmetric waveguide. (b) Distribution of power before and after the test up to $I = 2$ kA.

The damage sustained by the bar during the high current test is shown in figure 6.8 (b). Before the 2 kA test, the power distribution across the bar was very uniform around $P = 10.4$ W per emitter with a standard deviation of the mean $\sigma_P = 0.19$ W. At an equivalent current of $I = 10.8$ A, the single emitter emits $P = 10.8$ W. The small difference between the output powers from the emitters shows the good fabrication quality of the bar. After the test, four emitters of the bar have suffered catastrophic optical damage and the deviation in power of the working emitters has risen to $\sigma_P = 0.22$ W.

Figure 6.9 (a) shows the measurement results of the low temperature test of the bar and scaled single emitter using the thin, asymmetric waveguide design with $x = 25\%$ in the waveguide and 940 nm room temperature operation. The peak efficiency of the bar is reduced to 70% compared to the low Al-content designs. At a power of 1.5 kW efficiency is down to 59%. Roll over of power is delayed and starts at a current of $I \sim 1.2\,\text{kA}$. The main setback to output power are two kinks at $1.54\,\text{kA}$ and $1.95\,\text{kA}$, suggesting the failure of emitters on the bar at these currents. Despite the damaged emitters, the bar reaches $2\,\text{kW}$ output power with $\eta_E = 50\%$. Thermal resistance is $0.011\,\text{K/W}$ and $0.01\,\text{K/W}$ for the bar and single emitter, respectively. The significant increase in output power compared to the lower Al-content designs is attributed to the higher barriers around the active region suppressing power roll over at the high currents. At the same time, the higher series resistance due to the increased Al-content in the waveguide as well as due to the poor n-side metallization leads to the lowest efficiencies at the high currents of all designs in this iteration, despite the stronger drop in power of the low Al-content designs. Compared to scaled single emitter measurements, the PI curve of bar shows a marginally decreased threshold current as well as a slightly lower slope, both in accordance with the characteristic parameter results. At a scaled current of $1.18\,\text{kA}$, the single emitter experiences a sudden drop in power and at higher currents, the power current characteristic is not smooth. In the recorded spectrum shown in figure 6.10, a shift in emission wavelength is observed at the corresponding current of $31.9\,\text{A}$. Such a big negative shift in wavelength can be the result of a vertical mode hop, which is assumed here, causing output power to drop. Up to the point of the onset of the higher order mode, the lower series resistance and higher slope of the scaled single emitter (compared to the bar) result in a higher efficiency of 63% compared to 61.2%. After the optical mode hop, the efficiency of the single emitter is significantly reduced and does not reach the bar levels at an equivalent power.

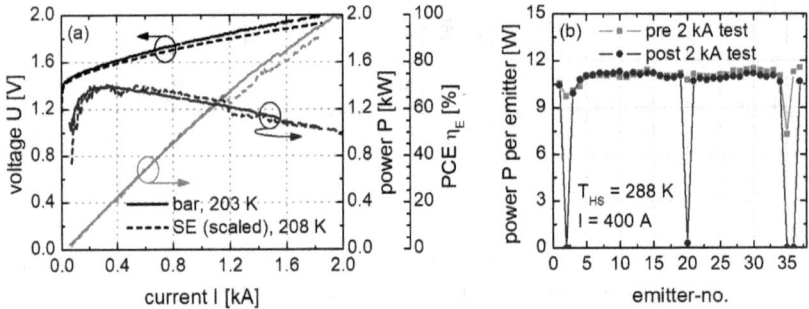

Figure 6.9: (a) Electro-optical characteristics of a bar at $T_{\text{HS}} = 203\,\text{K}$ and a scaled single emitter at $208\,\text{K}$ both with 25% Al-content in the thin, asymmetric waveguide. (b) Distribution of power prior and after the test up to $I = 2\,\text{kA}$.

The power distribution across the bar before and after the high current test is displayed in figure 6.9 (b). Going into the $2\,\text{kA}$ test, the different single emitter already had major deviations in output power. While emitters 1-4 have a slightly decreased performance, emitter 35 shows a significant drop in power. Overall, the standard deviation of the mean is $\sigma_P = 0.72\,\text{W}$. This explains the lower slope of the bar compared to the single emitters even at low currents. The high currents the bar was subjected to caused the failure of four emitters, two of which are the before mentioned emitters 2 and 35.

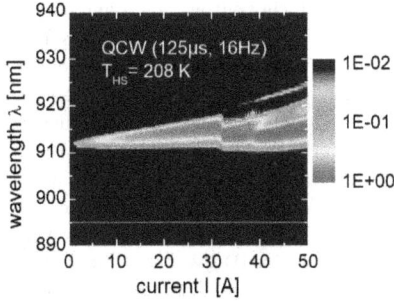

Figure 6.10: Recorded spectrum versus current for a single emitter with the thin, asymmetric waveguide with 25% Al-content.

The recorded spectrum of the single emitter is shown in figure 6.10. At a current of $I = 31.9\,A$, the central wavelength shifts from 915.2 nm to 914 nm as a higher order vertical mode reaches threshold and starts lasing. This parasitic mode can be suppressed in two ways. First, the vertical structure could be changed to inflict higher internal loss to this mode or decrease its confinement factor. These changes may also negatively impact the fundamental vertical mode which is undesirable. A second possibility is to increase the mirror loss for the higher order mode so it does not reach threshold. Without changing the reflectivity for the fundamental mode, this can be done by changing the thickness of the layers deposited on the front facet using coating tailored to suppress higher order vertical modes as reported in [Cru09.1].
The bar results are summarized in table 6.6.

Table 6.6: Summarized values from QCW bar measurements for different vertical structures from iteration 2 and iteration 1.

Al [%]	$P_{max.}$ [kW]	$U(2\,kA)$ [V]	$\lambda(1.5\,kW)$ [nm]	$\eta_{E,max.}$ [%]	$\eta_E(1.5\,kW)$ [%]	$\eta_E(P_{max.})$ [%]
7.5	1.88	1.77	936	77	62	53
15	1.95	1.82	942	75	62	53
25	2	2.02	915	70	59	50
7.5	*1.92*	*1.86*	*939*	*71*	*56*	*52*
4(DQW)	*2*	*1.81*	*941*	*69*	*61*	*55*

The bars using the thin, asymmetric waveguides with different Al-contents reach higher efficiencies at the targeted output power than the bars from iteration 1 using the Asloc designs with a SQW. Two of the three design reach maximum powers of $\sim 2\,kW$ at $T_{HS} = 203\,K$ even with emitters failing before the necessary current levels. For the $x = 7.5\%$ design, efficiency at all power levels is increased compared to iteration 1 due to a significantly decreased series resistance and less power saturation at higher currents. However, all bars suffer from multiple emitter failures during the test which is attributed

127

to the decreased spot size d/Γ, reducing output power and efficiency. Depending on the Al-content in the waveguide, the bars suffer to different degrees from power roll over at higher currents. While the $x = 7.5\%$ design is significantly affected by power saturation due to the low barriers around the active region, the highest Al-content structure is impaired by the high series resistance and the onset of lasing of a higher order vertical mode in the single emitter case. Consequently the design with 15% Al in the waveguide is the best option for high power, high efficiency operation, providing high enough barriers as well as a low enough series resistance. Comparing the best performing designs Asloc4DQW from iteration 1 and $x = 15\%$ from iteration 2, the achieved power levels as well as the voltage at 2 kA are similar, though the Asloc4DQW has a lower Al-content in the waveguide and is not affected by the degraded n-metallization or emitter failures. Peak efficiency is higher in the $x = 15\%$ design due to the lower number of QWs. At high currents, this negatively affects the power, though, as the power-current curve starts to roll over. Thus, at high powers, the efficiency values of the two designs are similar again, despite the lower series resistance of the $x = 15\%$ design.

The comparison with the single emitters sheds light on two unfavorable effects caused by the wafer process: an umpumped region in the middle of the single emitter cavity increases the threshold current and a faulty n-side metallization increases series resistance especially for the mounted bars of the $x = 25\%$ and 15% designs.

6.2.5 Estimation of loss mechanisms

In order to identify a path to even higher efficiency, the distribution of dissipated power, e.g. the fraction of electrical input power not being transformed into optical output power, across the different loss channels in the laser is analyzed following the approach in section 5.2.5 in the previous chapter. As the bars in this iteration suffered from failing or non-ideal emitters, the estimation of the magnitude of the different loss channels is again performed on the single emitters. This time, however, the single emitter width matches the emitter width on the bar and no further improvements to $P_{I_{th}}$ and P_{η_d} due to reduced current spreading are expected. The main efficiency limiting factors for the high power single emitter measurements at $T_{HS} = 208$ K are assessed and further alterations of the vertical structure are discussed analyzing the loss mechanisms at this temperature. The point of operation is chosen to be $P = 46$ W, as it is the highest power reached by all lasers of the three vertical structures and corresponds to 1.7 kW from a bar. In table 6.7 the values for efficiency, total waste heat and the contributions of the individual loss channels at P_{op} are listed for the three designs. Both $P_{joule,lin}$ as well as $P_{joule,\Delta U}$ are calculated and the difference between them is a measure for the bending of the UI-curve at higher currents.

Across all structures, the Joule heating due to the series resistance amounts to more than 50% of the total waste heat, even for the designs with low Al-content in the waveguide. There is, however, a strong difference in the magnitude of $P_{joule,lin}$ and $P_{joule,\Delta U}$ between the structures. The expected Joule heating as calculated from the linear series resistance is lowest for the $x = 7.5\%$ design, as R_S is the smallest of the three structures, followed by the 15% and lastly the 25% Al-content structure with almost twice as much Ohmic loss. The actual Joule heating calculated from the difference in turn on voltage and maximum voltage shows a quite different picture as the $x = 15\%$ design has the lowest $P_{\text{joule},\Delta U}$. While the lowest Al-content structure has a lower than expected Joule heating, it is no longer the design with the least Ohmic loss. Compared to the highest Al-content, measured Joule heating is only 16% lower, whereas a 45% decrease was expected from $P_{\text{joule,lin}}$. This means, that the assumption of a linear series resistance all the way up to the maximum current is not correct and that R_S extracted from the UI-curve at low currents decreases with increasing current. This is visible in a bending of the current-voltage characteristics in figures 6.7-6.9. In the graphs the bending of the UI-curve becomes more pronounced with higher Al-contents in the waveguides, which is also reflected in the increasing difference between $P_{\text{joule,lin}}$ and $P_{\text{joule},\Delta U}$ with increasing x in table 6.7. This behavior was already observed in the previous chapter, in which higher Al-contents in the waveguide led to a stronger bending of the voltage-current characteristic.

Table 6.7: Distribution of dissipated power at an optical output power of 46 W across loss channels for the three vertical designs with different Al-content in the waveguide, measured at $T_{\text{HS}} = 208\,\text{K}$ in single emitter configuration with 186 μm stripe width.

	$x = 7.5\%$	$x = 15\%$	$x = 25\%$
η_E [%]	56.9	62.4	53.2
P_{waste} [W]	35.6	28.4	41.2
$P_{I_{\text{th}}}$ [W]	1.0	1.0	1.1
P_{η_d} [W]	5.9	5.8	5.0
$P_{T_0 T_1}$ [W]	9.6	4.7	10.8
$P_{U_{\text{def}}}$ [W]	1.0	1.0	3.2
$P_{\text{joule,lin}}$ [W]	21.9	24.0	39.5
$P_{\text{joule},\Delta U}$ [W]	18.0	15.8	21.3

The higher Al-content structures do not only benefit more strongly from a decreasing series resistance from lower to higher currents, but also require less current to reach the power of 46 W. Since the current impacts P_{joule} quadratically (based on the formula for $P_{\text{joule,lin}}$), a higher threshold current, lower slope or higher thermal roll over of the PI-characteristics has a strong effect on the Joule heating. Comparing the $x = 7.5\%$ and 15% design in table 6.7, the dissipated power due to a lower slope and due to the threshold is the same, whereas $P_{T_0 T_1}$ is more than twice as high for the lower Al-content structure. The onset of the higher vertical order mode seen in the 25% Al-content structure impacts $P_{T_0 T_1}$, adding significantly to power saturation. The higher power saturation thus does not only increase the dissipated heat directly, but by increasing the necessary current for a certain power in turn raises P_{joule}. Based on the findings in chapter 5 that the higher roll over is caused by the lower barriers around the quantum well due to the lower Al-content in the waveguide, the lower series resistance (at low currents) is hence at the expense of a higher I_{op} and and a higher $P_{T_0 T_1}$.

The analysis of all the loss mechanisms also reveals, that the threshold and defect voltage at turn on do not significantly decrease efficiency. Together they only cause a maximum of 10% of the dissipated power in the case of the $x = 25\%$ design and contribute as little as 6% for the 15% Al-content structure. The absolute penalty due to a non ideal slope is similar between the designs which is obvious as internal loss and internal efficiency at this low T_{HS} are the same. As P_{waste} decreases, however, the relative contribution increases, as can be seen in the $x = 15\%$ case in which P_{η_d} contributes 20% to the total dissipated heat.

While the thin, asymmetric waveguide with a low Al-content successfully decreased series resistance, it only causes a benefit at lower currents. At the high powers and currents, however, deterioration of the PI-curve increases operation current and thus Ohmic heating. A possible solution to limit power roll over is the implementation of a higher number of wells, e.g. a double quantum well. This proved to significantly decrease roll over in low Al-content structures in chapter 5. The higher threshold current from a DQW should not be deteriorating performance significantly, as the energy used to establish threshold is the lowest value of all loss mechanisms for the SQW designs. Here, a tradeoff between a higher $P_{I_{th}}$ for a lower $P_{T_0 T_1}$ and thus a lower I_{op} is favorable. However, the higher number of carriers in the quantum wells (for the DQW compared to the SQW) can lead to higher optical absorption and thus deteriorate P_{η_d} as reported in [Kau16].

6.3 Conclusions from iteration 2

In this chapter, the use of thin, asymmetric waveguides combined with low Al-contents was investigated targeting the p-waveguide as the main contributor to series resistance. Starting with short loop processed devices, three designs with different Al-content in the waveguide were assessed. Compared to structures from iteration 1, series resistance (excluding substrate and contact contributions) was successfully reduced by $> 44\%$ for designs with the same Al-content, while internal loss stayed roughly constant. Fully processed bars with the asymmetric, thin waveguide design reach $\sim 2\,\text{kW}$ in output power at $T_{HS} = 203\,\text{K}$ with an efficiency of 53%, deteriorated by emitter failures and a faulty n-side metallization. At the targeted power of $1.5\,\text{kW}$, the designs reached higher efficiencies of 62% than achieved in iteration 1. Furthermore, record peak efficiencies of 77% from a $1\,\text{cm}$ high fill factor bar were measured. While showing the lowest voltage ever observed in this work at a maximum current of $2\,\text{kA}$, bars with the vertical design with the lowest Al-content in the waveguide suffered from significant roll over limiting power and thus efficiency at this high current. A thorough analysis and comparison of the loss channels at high powers revealed that Joule heating was still the main contributor to waste heat even in the low Al-content, thin asymmetric waveguide structures. Actually, while the lowest Al-content structures showed the lowest series resistance, they suffered from higher Ohmic losses than a higher Al-content design, which was caused by the increased current needed for the high power levels due to the increased roll over and a stronger bending of the UI-curve for higher Al-contents. A solution to benefit from the low resistance as well as a low power saturation would be to increase the number of quantum wells, as successfully demonstrated in iteration 1. This would be feasible as the power used for establishing threshold in all designs was the lowest of all loss mechanisms. However, the higher number of carriers present in a higher number of wells could potentially increase optical absorption. Overall, thin asymmetric low Al-content waveguides proved to significantly reduce series resistance and a path for resolving the remaining challenges of increased power saturation has been presented.

Chapter 7

Summary and Outlook

Summary

Within this work, high power diode laser bars optimized for highly efficient low temperature operation have been developed. Progress in bar performance relied on the detailed assessment of single emitters, which were evaluated on their temperature dependent electro-optical behavior. Three design iterations were performed analyzing the impact of the epitaxial vertical structure on the efficiency limiting factors at different temperatures. After a benchmarking iteration, the development of the DL bars was done in three consecutive steps:

- The epitaxial vertical structure was modeled using a vertical mode solver to achieve low resistance, low loss structures.

- Single emitters of the vertical designs were fabricated in a short loop process and measured up to low currents at temperatures down to $200\,\mathrm{K}$.

- The most promising designs to reach high output powers and efficiencies were processed into single emitters and high fill factor bars, which were subsequently experimentally assessed up to high currents.

For the theoretical evaluation of vertical laser designs, the simulation tool QIP [Wen90] was used. Amongst other parameters, it computes the expected values for internal loss, optical confinement factor and series resistance from the stack of epitaxial bulk semiconductor layers. Furthermore, the energy of the bound quantum well ground state as well as the conduction and valence band energy levels is calculated. The difference between the band edge in the waveguide and quantum well energy levels ΔE, e.g. the barrier height around the quantum well, is used as a measure of the carrier confinement in the well. Subsequently, theoretically assessed designs were processed into broad area single emitters omitting process steps necessary for high power devices in turn for fast process times. The SE were tested to low currents at various temperatures T to obtain the temperature dependence of the characteristic parameters external differential efficiency η_d, threshold current I_th and series resistance R_S for the different vertical designs. For one specific design, diodes of three different lengths were also tested, so that the T-dependence of the internal parameters internal differential efficiency η_i, internal loss α_i, transparency current density J_T and modal gain Γg_0 could be obtained. For a full laser process, designs were developed using extrapolations based on the theoretical evaluation and low current findings. The performance of the resulting high power single emitters was checked against the behavior of the equivalent short loop processed diodes and measurements up to high currents provided an upper limit for the bar performance. Finally, high power, high efficiency diode laser bars optimized for low operation temperatures were driven up to the maximum available current of 2kA. The experimental findings and technical advancements achieved over the course of this work are subsequently presented.

Experimental findings

The analysis of the T-dependence of the internal parameters revealed, that η_i tends to 100% for lower T, with the maximum measured value of 96% determined at 218 K, the lowest assessed temperature for the short loop processed diodes. The strong increase of η_i at lower T is attributed to the decrease of leakage currents and non-radiative recombination channels. No clear change in internal loss is observed at reduced T, so that the increase in η_i is assumed to be the main reason behind the previously reported increases in η_d in [Cru06]. Similarly, the decrease of J_T at lower T is also attributed to reduced carrier leakage and non-radiative recombination, explaining the observed lower thresholds at low T in [Zha94, Cru06, Mai08]. In addition, new understanding on the low T behavior of diode lasers was developed during the low current measurements and pertained to the increase of R_S at lower T. This observation is contrary to the T-dependence of resistance in semiconductor bulk layers in the assessed temperature regime [Sin12, Sot00]. Analyzing single emitters with various vertical structures using different carrier confinements in the quantum well, a clear dependence of series resistance on effective barrier height around the quantum well was found, as depicted in figure 7.1.

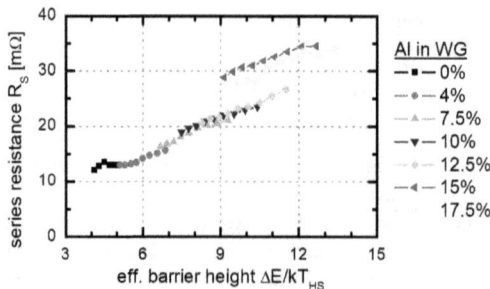

Figure 7.1: Measured series resistance plotted against effective barrier height $\Delta E/k_B T_{HS}$ for designs using different Al-contents in the waveguides.

The excess resistance was attributed to the degraded transport of carriers across the QW barriers. Other structure related factors could be excluded, including the transition from the cladding to the waveguide, which is a common problem in InP lasers [Lei10]. Initial trials showed, that the excess resistance could be reduced using a slight p-doping of the active region. For a diode laser bar, the benefit of the doping was seen in a reduced voltage while output power up to the assessed 1 kA was not affected. Still, R_S remained higher than expected from bulk layer calculations, but a path to reduce the excess resistance was shown and the result directly confirms the resistance contribution of the active region. Realized 1 cm wide QCW bars with high fill factors (69%) and a cavity length of $L = 4$ mm achieved record output powers of 2 kW at an operation temperature of 203 K [Fre15] as shown in figure 7.2, limited by the available current of 2 kA. The power-current characteristic shows little roll-over as power saturation effects at higher currents are suppressed by the use of a double quantum well. Bars using a similar vertical structures with the same barrier height, but a single quantum well showed significant roll over at currents above 1.2 kA. At the same time, no indication of additional free carrier absorption due to the higher number of carriers in the active region is observed in the bar using the double quantum well.

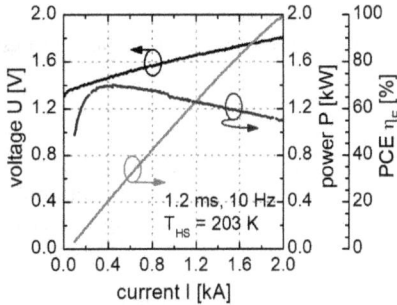

Figure 7.2: Power, voltage and efficiency plotted against the current for the high fill factor bar reaching 2 kW in output power.

The low series resistance of the bar was achieved by a low-temperature optimized vertical structure using low Al-content $Al_{0.04}Ga_{0.96}As$ waveguides. This resulted in high efficiencies at high powers, namely 55% at the maximum output power. At the target specifications (≥ 1.5 kW) efficiencies remained above 60%, which is also in line with the set goals. Record peak efficiencies of 77% were reached from bars using a vertical design with a thin p-waveguide and a low Al-content in the waveguides of $x = 7.5\%$. At a power of 1.5 kW, efficiency was still 62% and thus within the target specification of this work.

Technical advancements

Characterizing QCW bars at low temperatures required the careful design and set up of a high power, low temperature measurement station. The final measurement system allowed the QCW operation of bars and single emitters as well as CW operation of SE over a wide range of temperatures down to 200 K. The challenge of accurately determining QCW power levels was addressed by precisely analyzing the optical pulse shape and establishing a power calibration referenced against in-house and national power standards. Aside from the power measurement of QCW bars, determining the voltage drop across the laser diode posed a challenge, as a 4-terminal measurement is not possible and package resistance is superimposed. For this work, low stress mounting schemes using thick Cu-blocks rather than bond wires or Cu-foils were developed for conduction cooled bar packages (CCP), reducing package resistance to $\sim 10\mu\Omega$. Using these assemblies, the voltage and thus the diode's series resistance as well as its power conversion efficiency could be accurately measured. The new, robust mounting scheme facilitates the implementation of assembled bars into solid-state laser systems.

Before the beginning of this work, short loop processed devices were only used as uncoated, unmounted bars to determine the internal and other laser parameters in short pulse mode at room temperature from length-dependent measurements. The need to accurately measure the voltage and the laser performance of many devices and structures with short cycle times and low fabrication cost over a wider range of temperatures led to a new p-side up mounting scheme of short loop processed and facet coated single emitters using Al-bond wires to contact the thick Al-metal on the p-side. This technique is now routinely used at the FBH to more thoroughly test experimental epitaxial designs which have been short-loop processed into single emitters before a full laser process.

Outlook

Reducing the excess resistance at low currents remains a challenge on the way to higher efficiencies at high powers. The discrepancy between measured and theoretically expected series resistance as well as the dependence of excess resistance on the barrier height around the active region is currently being studied. First positive results with lower excess series resistance have been achieved by doping the active region and further investigations will have to be conducted to gauge the upper feasible limit of this effect, e.g. at which point free carrier absorption due to the additional doping will deteriorate optical performance. Tailoring the active region to mitigate the degraded transport of carriers across the barriers will also include studies of the adjacent layers by introducing various graded index profiles. Again, the impact on the optical behavior especially at high currents will have to be considered. Ongoing studies at the FBH treat the mitigation of power saturation effects at high currents [Wan10, Kau16], which have also been observed in this work. While the use of double quantum wells promises a more stable high current operation, other concepts to decrease leakage will also have to be explored, for example the use of additional barrier layers [Asr11].

Technologically, the implementation of monolithic gratings in the laser structure spectrally stabilizes the emitted light and allows operation on a narrow spectral line. Introducing these gratings into the developed laser structures without inferring additional optical losses or electrical resistance as shown in [Sch10] would help confine the laser spectrum to one of the narrow desired pumping wavelengths and increase the pumping efficiency [Ert11]. Furthermore, the transfer of the results to other wavelengths would enable pumping of different solid-state media. Finally, the high efficiencies and the low dissipated heat in the presented QCW bars permit to bundle multiple bars into diode laser stacks and thus increase the power density of the device. The construction of high density stacks as suggested in [Fee11] would reduce the cost of the device by removing individual heatsinks and required lenses for beam control.

Appendix

Full wafer processing

The processing of epitaxially grown wafers provides lateral current and optic confinement as well as establish the ability to electrically connect the resulting laser bars (containing the single emitters). Each wafer is treated with a p-side and an n-side process and consists of multiple process steps.These steps are tailored to the epitaxial structure used.

On the p-side, helium-implantation is first used to define contact windows of single emitters with stripe width w by damaging the crystal structure of the surrounding wafer surface. Specifically, mid-band defects are created which render the highly doped GaAs contact layer electrically insulating defining the contact areas. As a next step, shallow index trenches running parallel to the contact window are formed by reactive ion etching (RIE). These trenches, situated several micrometers next to the contact area of the single emitters serve as weak lateral index guides. Furthermore, etching of deep trenches at a bigger distance of the contact window provide electrical and optical isolation between the single emitters, defining the chip (not stripe) width. Subsequently, a SIN_x isolator layer is applied on the whole wafer surface using plasma-enhanced chemical vapor deposition. The SIN_x covering the contact windows is then removed by RIE. Providing electrical connectivity, the p-side is then metallized, depositing multiple layers of Ti, Pt and Au on top of the contact window. Finally, a 3 µm thick Au layer for uniform current and heat spreading is added to each stripe by galvanization.

The n-side process entails as a first step the thinning of the substrate to a thickness of ~ 150 µm. The resulting damages on the surface are removed by a 20 µm relaxation etch. Subsequently, an ohmic contact is established by depositing Ni-Au-Ge as back side metal. Finally, a second metallization of Ti, Pt and Au enables wire bonding of the n-side and concludes the processing of the wafer.

Short loop processing

A full laser process (FP) is specifically adjusted for the vertical epitaxial design and the intended lateral laser structure to achieve the best result possible for a certain vertical epitaxial design. For a fair assessment of the vertical epitaxial design itself a standardized lateral laser structure and process is needed. Moreover, the use of thick Au on the p-side raises the monetary cost of a FP further and is not necessary when comparing devices that were processed identically.

A simple solution to test vertical epitaxial designs was developed at the FBH, providing short process times ($\sim$1 week) and using Al as a p-side contact. This method is termed short loop process (SLP). The short cycle times enable quick experimental assessment of the vertical structures and are gained by a standardization of the process, independent of epitaxial design, combined with omitting process steps used in the FP. First, a fixed set of stripe widths and lengths is used for each wafer. Second, the p-side metallization is applied *before* the implantation, serving as a mask for the implantation. Furthermore, the isolator is omitted. Lastly, a uniform current distribution across the length of the stripe is provided by a 4 µm thick Al-layer.

Originally the SLP was used to determine the internal parameters of the vertical laser structure and no facet coating or mounting of laser diodes was intended. For example, the p-side Al complicates contacting and prohibits p-down mounting. However, with the limitations in mind, the SLP can still be a useful tool to gain fully assembled single emitters in short cycle times.

List of Figures

List of Tables

Bibliography

[Li00] L. Li, *The advances and characteristics of high-power diode laser materials processing*, Optics and Lasers in Engineering **34(4)**, 231 (2000).

[Sch00] W. Schulz, and R. Poprawe, *Manufacturing with novel high-power diode lasers*, IEEE Journal of Selected Topics in Quantum Electronics **6(4)**, 696 (2000).

[Nas14] H. Nasim, and Y. Jamil, *Diode lasers: From laboratory to industry*, Optics and Laser Technology **56**, 211 (2014).

[Sch10] C.M. Schultz, P. Crump, H. Wenzel, O. Brox, A. Maaßdorf, G. Erbert, and G. Tränkle, *11W broad area 976 nm DFB lasers with 58% power conversion efficiency*, Electronic Letters **46(8)**, 580 (2010).

[Mas11] P. D. Mason, K. Ertel, S. Banerjee, P. J. Phillips, C. Hernandez-Gomez, and J.L. Collier, *Optimised design for a 1 kJ diode-pumped solid-state laser system*, Proc. SPIE **8080**, 80801X (2011).

[Ert11] K. Ertel, S. Banerjee, P. D. Mason, P. J. Phillips, M. Siebold, C. Hernandez-Gomez, and J. C. Collier, *Optimising the efficiency of pulsed diode pumped Yb:YAG laser amplifiers for ns pulse generation*, Optics Express **19(27)**, 26610 (2011).

[Fan07] T. Y. Fan, D. J. Ripin, R. L. Aggarwal, J. R. Ochoa, B. Chann, M. Tilleman, and J. Spitzberg, *Cryogenic Yb3+-doped solid-state lasers*, IEEE Journal of Selected Topics in Quantum Electronics **13**, 448 (2007).

[Mil04] G. Miller, E. Moses, and C. Wuest, *The national ignition facility*, Optical Engineering **43**, 2841 (2004).

[Der11] R. Deri et al., *Semiconductor Laser Diode Pumps for Inertial Fusion Energy Lasers*, LLNL-TR-465931, Lawrence Livermore National Laboratory (2011).

[Mos11] E. I. Moses, *The National Ignition Facility and goals of near-term laser fusion energy*, CLEO Europe 2011, paper CA4_1 (2011).

[Mas11.2] P. Mason, S. Banerjee, K. Ertel, C. Hernandez-Gomez, J. Phillips, and J. Collier, *Multi-slab architecture for kJ-class DPSSL system*, CLEO Europe 2011, paper CA4_2 (2011).

[NIF16] Anon., *National Ignition Facility - User Guide 2016*, LLNL-TM-681123_P2103437_WO15466_NUG, Lawrence Livermore National Laboratory (2016).

[Kan05] M. Kanskar, T. Earles, T.J. Goodnough, E. Stiers, D. Botez, and L. J. Mawst, *73% CW power conversion efficiency at 50W from 970nm diode laser bars*, Electronic Letters **41(5)**, 245 (2005).

[Cru05] P. Crump, J. Wang, T. Crum, S. Das, M. DeVito, W. Dong, J. Farmer, Y. Feng, M. Grimshaw, D. Wise, and S. Zhang, *> 360 W and > 70% efficient GaAs-based diode lasers*, Proc. SPIE **5711** (2005).

[Cru13] P. Crump, G. Erbert, H. Wenzel, C. Frevert, C.M. Schultz, K.-H. Hasler, R. Staske, B. Sumpf, A. Maaßdorf, F. Bugge, S. Knigge, and G. Tränkle, *Efficient high-power laser diodes*, IEEE Journal of Selected Topics in Quantum Electronics **19**, 1501211 (2013).

[Lau12] C. Lauer, H. König, G. Grönninger, S. Hein, A. Gomez-Iglesias, M. Furitsch, J. Maric, H. Kissel, P. Wolf, J. Biesenbach, and U. Strauss, *Advances in performance and beam quality of 9xx-nm laser diodes tailored for efficient fiber coupling*, Proc. SPIE **8241**, 824111 (2012).

[Cru09] P. Crump, G. Blume, K. Paschke, R. Staske, A. Pietrzak, U. Zeimer, S. Einfeldt, A. Ginolas, F. Bugge, K. Häusler, P. Ressel, H. Wenzel, and G. Erbert, *20W continuous wave reliable operation of 980nm broad- area single emitter diode lasers with an aperture of 96µm*, Proc. SPIE **7198**, 719814 (2009).

[Cru06] P. Crump, M. Grimshaw, J. Wang, W. Dong, S. Zhang, S. Das, J. Farmer, M. DeVito, L. S. Meng, and J. K. Brasseur, *85% Power Conversion Efficiency 975-nm Broad Area Diode Lasers at -50 °C, 76 % at 10 °C*, presented at the Quantum Electron. Laser Sci. Conf., Long Beach, CA, USA (2006).

[Mai08] M. A. Maiorov, and I. E. Trofimov, *Diode laser pumping sources for cryogenically cooled solid-state lasers*, Proc. SPIE **6952**, 695209 (2008).

[Lei10] P. O. Leisher, W. Dong, M. P. Grimshaw, M. J. DeFranza, M. A. Dubinskii, and S. G. Patterson, *Mitigation of Voltage Defect for High-Efficiency InP Diode Lasers Operating at Cryogenic Temperatures*, IEEE Photonics Technology Letters **22(24)**, 1829 (2010).

[Die00] R. Diehl (Ed.), *High-power diode lasers fundamentals*, Topics in Applied Physics **vol. 78**, Springer Berlin Heidelberg (2000).

[Dei08] E. Deichsel, D. Schröder, J. Meusel, R. Hülsewede, J. Sebastian, S. Ludwig, and P. Hennig, *Highly reliable qcw laser bars and stacks*, Proc. SPIE **6876**, 68760K (2008).

[Ber11] Y. Berk, Y. Karni, G. Klumel, Y. Openhaim, S. Cohen, and D. Yanson, *Scaleable multi-format QCW pump stacks based on 200W laser diode bars and mini bars at 808nm and 940nm*, Proc. SPIE **7918**, 79180W (2011).

[Koh12] A. Kohl, T. Fillardet, A. Laugustin, and O. Rabot, *Ultra high brightness laser diode arrays for pumping of compact solid state lasers and direct applications*, Proc. SPIE **8546**, 854608 (2012).

[Luc15] A. Lucianetti, J. Pilar, A. Pranovich, M. Divoky, T. Mocek, K. Ertel, H. Jelinkov, P. Crump, C. Frevert, R. Staske, G. Erbert, and G. Tränkle, *Assessment of high-power kW-class single-diode bars for use in highly efficient pulsed solid-state laser systems*, Proc. SPIE **9348**, 934811 (2015).

[Sch07] D. Schröder, J. Meusel, P. Henning, D. Lorenzen, M. Schröder, R. Hülsewede, and J. Sebastian, *Increased power of broad area lasers (808nm / 980nm) and applicability to 10mm-bars with up to 1000Watt QCW*, Proc. SPIE **6456**, 64560N (2007).

[Li07] H. X. Li, I. Chyr, X. Jin, F. Reinhardt, T. Towe, D. Brown, T. Nguyen, M. Berube, T. Truchan, D. Hu, R. Miller, R. Srinivasan, T. Crum, E. Wolak, R. Bullock, J. Mott, and J. Harrison, *> 700 W continuous-wave output power from a single laser diode bar*, Electronic Letters **43**, 25 (2007).

[Li08] H. X. Li, F. Reinhardt, I. Chyr, X. Jin, K. Kuppuswamy, T. Towe, D. Brown, O. Romero, D. Liu, R. Miller, T. Nguyen, T. Crum, T. Truchan, E. Wolak, J. Mott, and J. Harrison, *High-efficiency, high-power diode laser chips, bars, and stacks*, Proc. SPIE **6876**, 68760G (2008).

[Kna11] M.T. Knapczyk, J.H. Jacob, H. Eppich, A.K. Chin, K.D. Lang, J.T. Vignati, and R. H. Chin, *70% efficient near 1kW single 1-cm laser-diode bar at 20 °C*, Proc. SPIE **7918**, 79180F (2011).

[Kis13] H. Kissel, W. Fassbender, J. Lotz, K. Alegria, T. Koenning, D. Stapleton, S. Patterson, and J. Biesenbach, *Reliable QCW diode laser arrays for operation with high duty cycles*, Proc. SPIE **8605**, 86050V (2013).

[Vet14] T. Vethake, *High energy pulse stack HEC-DPSSL*, presented at the 8th international HEC-DPSSL workshop, Oxford, UK (2014).

[Wol14] M. Wölz, *Increasing the power density towards IFE requirements: high-power laser diode bars*, presented at the 8th international HEC-DPSSL workshop, Oxford, UK (2014).

[Che14] Z. Chen, J. Bai, W. Dong, X. Guan, S. Zhang, S. Elim, L. Bao, M. Grimshaw, M. Devito, and M. Kanskar, *High Power and High Efficiency kW 88x-nm Multi-Junction Pulsed Diode Laser Bars and Arrays*, Proc. SPIE **8965**, 896514-1 (2014).

[Pie15] A. Pietrzak, M. Wölz, R. Hülsewede, M. Zorn, O. Hirsekorn, J. Meusel, A. Kindsvater, M. Schröder, V. Blümel, and J. Sebastian, *Heading to 1 kW levels with laser bars of high-efficiency and emission wavelength around 880 nm and 940 nm*, Proc. SPIE **9348**, 93480E (2015).

[Wes13] C. Wessling, O. Rübenach, S. Hambücker, V. Sinhoff, S. Banerjeea, K. Ertel, and P. Mason, *Efficient pumping of inertial fusion energy lasers*, Proc. SPIE **8602**, 86020I (2013).

[Cru14] P. Crump, C. Frevert, H. Hösler, F. Bugge, S. Knigge, W. Pittroff, G. Erbert, and G. Tränkle, *Cryogenic ultra-high power infrared diode laser bars*, Proc. SPIE **9002**, 90021I (2014).

[Fri12] J. Fricke, H. Wenzel, F. Bugge, O. Brox, A. Ginolas, W. John, P. Ressel, L. Weixelbaum, and G. Erbert, *High-power distributed feedback lasers with surface gratings*, IEEE Photonics Technology Letters **24(16)**, 1443 (2012).

[Col98] L. A. Coldren, M. L. Mashanovitch, and S. W. Corzine, *Diode Lasers and Photonic Integrated Circuits*, Wiley 2nd ed. (2012).

[Sze01] S. M. Sze, *Physics of Semiconductor Devices*, Wiley New York, 2nd ed. (2001).

[Ada93] S. Adachi, *Properties of Aluminium Gallium Arsenide*, INSPEC (1993).

[DeT93] T. A. DeTemple, and C. M. Herzinger, *On the semiconductor laser logarithmic gain-current density relation*, IEEE Journal of Quantum Electronics **29(5)**, 1246 (1993).

[Kau18] T. Kaul, G. Erbert, A. Maaßdorf, S. Knigge, and P. Crump, *Suppressed power saturation due to optimized optical confinement in 9xx nm high-power diode lasers that use extreme double asymmetric vertical designs*, Semiconductor Science and Technology **33(3)**, 035005 (2018).

[Wol70] C. M. Wolfe, G. E. Stillman, and W. T. Lindley, *Electron Mobility in High-Purity GaAs*, Journal of Applied Physics **41**, 3088 (1970).

[Sin12] M. Sintschuk, *Bestimmung der elektro-optischen Eigenschaften von* $Al_xGa_{1-x}As$ *in Abhängigkeit von Zusammensetzung und Dotierung*, bachelor thesis at Beuth Hochschule für Technik Berlin, Fachbereich II Mathematik - Physik - Chemie (2012).

[Fre16] C. Frevert, F. Bugge, S. Knigge, A. Ginolas, G. Erbert, and P. Crump, *940 nm QCW laser bars with 70% efficiency at 1 kW output power at 203 K: analysis of remaining limits and path to higher efficiency and power at 200 K and 300 K*, Proc. SPIE **9733**, 97330L (2016).

[Has14] K. H. Hasler, H. Wenzel, P. Crump, S. Knigge, A. Maaßdorf, R. Platz, R. Staske, and G. Erbert, *Comparative theoretical and experimental studies of two designs of high-power diode lasers*, Semiconductor Science and Technology **29 (4)**, 045010 (2014).

[Res05] P. Ressel, G. Erbert, U. Zeimer, K. Häusler, G. Beister, B. Sumpf, A. Klehr, and G. Tränkle, *Novel passivation process for the mirror facets of high-power semiconductor diode lasers*, IEEE Photonics Technology Letters **17**, 962 (2005).

[Wen00] H. Wenzel, G. Erbert, A. Knauer, A. Oster, K. Vogel and G. Tränkle, *Influence of current spreading on the transparency current density of quantum-well lasers*, Semiconductor Science and Technology **15(6)**, 557 (2000).

[Cru13.1] P. Crump, C. Frevert, H. Wenzel, F. Bugge, S. Knigge, G. Erbert, and G. Tränkle, *Cryolaser: innovative cryogenic diode laser bars optimized for emerging ultra-high power laser applications*, Conference on Lasers and Electro Optics, San Jose, USA, paper **JW1J.2** (2013).

[Bot99] D. Botez, *Design considerations and analytical approximations for high continuous-wave power, broad-waveguide diode lasers*, Applied Physics Letters **74.21**, p. 3102 (1999).

[Sot00] M. Sotoodeh, A.H. Khalid, and A.A. Rezazadeh, *Empirical low-field mobility model for III-V compounds applicable in device simulation codes*, Journal of Applied Physics **87.6**, p. 2890-2900 (2000).

[Pet07] M. Peters, V. Rossin, M. Everett, and E. Zucker, *High power, high efficiency laser diodes at JDSU*, Proc. SPIE **6456**, 64560G-5 (2007).

[Wen90] H. Wenzel, and H.-J. Wünsche, *A Model for the Calculation of the Threshold Current of SCH-MQW-SAS Lasers*, physica status solidi a **120.2**, p. 661-673 (1990).

[Pie09] A. Pietrzak, P. Crump, H. Wenzel, R. Staske, G. Erbert, and G. Tränkle, *55 W peak power from 1100 nm wavelength 60 μm broad-area laser diodes enabled by reduced carrier accumulation in the waveguide*, Semiconductor Science and Technology **24**, 035020 (2009).

[Erb00] G. Erbert, *High-power diode lasers fundamentals*, Topics in Applied Physics **vol. 78**, pp. 181, ed. R. Diehl (2000).

[Kau16] T. Kaul, G. Erbert, R. Platz, A. Maaßdorf, S. Knigge, and P. Crump, *Studies of limitations to peak power and efficiency in diode lasers using extreme-double-asymmetric vertical designs*, 25th International Semiconductor Laser Conference **WD4** (2016).

[Wen10] H. Wenzel, P. Crump, A. Pietrzak, X. Wang, G. Erbert, and G. Tränkle, *Theoretical and experimental investigations of the limits to the maximum output power of laser diodes*, New Journal of Physics **12**, 085007 (2010).

[Bou93] D. P. Bour, D. W. Treat, R. L. Thornton, R. S. Geels, and D. F. Welch, *Drift leakage current in AlGaInP quantum-well lasers*, IEEE Journal of Quantum Electronics **29**, 1337 (1993).

[Mor91] S. Morin, B. Deveaud, F. Clerot, K. Fujiwara, and K. Mitsunaga, *Capture of Photoexcited Carriers in a Single Quantum Well with Different Confinement Structures*, IEEE Journal of Quantum Electronics **27.6**, 1669 (1991).

[Spr10] M. Spreemann, B. Eppich, F. Schneider, H. Wenzel, and G. Erbert, *Modal behavior, spatial coherence, and beam quality of a high-power gain-guided laser array*, IEEE Journal of Quantum Electronics **46(11)**, 1619 (2010).

[Wan10] X. Wang, P. Crump, H. Wenzel, A. Liero, T. Hoffmann, A. Pietrzak, C.M. Schultz, A. Klehr, A. Ginolas, S. Einfeldt, F. Bugge, G. Erbert, and G. Tränkle, *Root-Cause Analysis of Peak Power Saturation in Pulse-Pumped 1100 nm Broad Area Single Emitter Diode Lasers*, IEEE Journal of Quantum Electronics **46(5)**, 658 (2010).

[Has17] K. H. Hasler, C. Frevert, P. Crump, G. Erbert, and H. Wenzel, *Numerical study of high-power semiconductor lasers for operation at sub-zero temperatures*, Semiconductor Science and Technology **32(4)**, 045004 (2017).

[Bul05] S. Bull, J. W. Tomm, M. Oudart, J. Nagle, C. Scholz, K. Boucke, I. Harrison, and E. C. Larkins, *By-emitter degradation analysis of high-power laser bars*, Journal of Applied Physics **98**, 063101 (2005).

[Cru04] P.A. Crump, T.R. Crum, M. DeVito, J. Farmer, M. Grimshaw, Z. Huang, S. A. Igl, S. Macomber, P. Thiagarajan, and D. Wise, *High efficiency, high power 808-nm laser array and stacked arrays optimized for elevated temperature operation*, Proc. SPIE **5336**, p. 145 (2004).

[Fre15] C. Frevert, P. Crump, F. Bugge, S. Knigge, A. Ginolas, and G. Erbert, *Low-temperature Optimized 940 nm Diode Laser Bars with 1.98 kW Peak Power at 203 K*, Conference on Lasers and Electro Optics, San Jose, USA, paper **SM3F.8** (2015).

[Epp13] P. W. Epperlein, *Semiconductor Laser Engineering, Reliability and Diagnostics: A Practical Approach to High Power and Single Mode Devices*, Wiley (2013).

[Cru09.1] P. Crump, H. Wenzel, P. Ressel, G. Erbert, and G. Tränkle, *Multiple vertical mode high power 975-nm diode lasers restricted to single vertical optical mode operation through use of optical facet coatings*, Electronic Letters **45**, 51 (2009).

[Zha94] B. Zhao, T. R. Chen, L. E. Eng, Y. H. Zhuang, A. Shakouri, and A. Yariv, *Sub-100 µA current operation of strained InGaAs quantum well lasers at low temperatures*, Applied Physics Letters **65**, 1805 (1994).

[Asr11] L. V. Asryan, N. V. Kryzhanovskaya, M. V. Maximov, A. Y. Egorov, and A. E. Zhukov, *Bandedge-engineered quantum well laser*, Semiconductor Science and Technology **26**, 055025 (2011).

[Fee11] R. Feeler, J. Junghans, and E. Stephens, *Low-Cost Diode Arrays for the LIFE Project*, Proc. SPIE **7916**, 791608-1 (2011).

Innovationen mit Mikrowellen und Licht
Forschungsberichte aus dem Ferdinand-Braun-Institut, Leibniz-Institut für Höchstfrequenztechnik

Herausgeber: Prof. Dr. G. Tränkle, Prof. Dr.-Ing. W. Heinrich

Cuvillier Verlag
Internationaler wissenschaftlicher Fachverlag

Innovationen mit Mikrowellen und Licht
Forschungsberichte aus dem Ferdinand-Braun-Institut, Leibniz-Institut für Höchstfrequenztechnik

Herausgeber: Prof. Dr. G. Tränkle, Prof. Dr.-Ing. W. Heinrich

Innovationen mit Mikrowellen und Licht
Forschungsberichte aus dem Ferdinand-Braun-Institut, Leibniz-Institut für Höchstfrequenztechnik

Herausgeber: Prof. Dr. G. Tränkle, Prof. Dr.-Ing. W. Heinrich

Band 21:   **Agnietzka Pietrzak**
Realization of High Power Diode Lasers with Extremely Narrow
Vertical Divergence
ISBN: 978-3-95404-066-7, 27,40 EUR, 144 Seiten

Band 22:   **Eldad Bahat-Treidel**
GaN-based HEMTs for High Voltage Operation
Design, Technology and Characterization
ISBN: 978-3-95404-094-0, 41,10 EUR, 220 Seiten

Band 23:   **Ponky Ivo**
AlGaN/GaN HEMTs Reliability:
Degradation Modes and Anslysis
ISBN: 978-3-95404-259-3, 23,55 EUR, 132 Seiten

Band 24:   **Stefan Spießberger**
Compact Semiconductor-Based Laser Sources
with Narrow Linewidth and High Output Power
ISBN: 978-3-95404-261-6, 24,15 EUR, 140 Seiten

Band 25:   **Silvio Kühn**
Mikrowellenoszillatoren für die Erzeugung von atmosphärischen
Mikroplasmen
ISBN: 978-3-95404-378-1, 21,85 EUR, 112 Seiten

Band 26:   **Sven Schwertfeger**
Experimentelle Untersuchung der Modensynchronisation in Multisegment-
Laserdioden zur Erzeugung kurzer optischer Pulse bei einer Wellenlänge
von 920 nm
ISBN: 978-3-95404-471-9, 29,45 EUR, 150 Seiten

Band 27:   **Christoph Matthias Schultz**
Analysis and mitigation of the factors limiting the effiency of high power
distributed feedback diode lasers
ISBN: 978-3-95404-521-1, 68,40 EUR, 388 Seiten

Band 28:   **Luca Redaelli**
Design and fabrication of GaN-based laser diodes for single-mode
and narrow-linewidth applications
ISBN: 978-3-95404-586-0, 29,70 EUR, 176 Seiten

Band 29:   **Martin Spreemann**
Resonatorkonzepte für Hochleistungs-Diodenlaser
mit ausgedehnten lateralen Dimensionen
ISBN: 978-3-95404-628-7, 25,15 EUR, 128 Seiten

Innovationen mit Mikrowellen und Licht
Forschungsberichte aus dem Ferdinand-Braun-Institut, Leibniz-Institut für Höchstfrequenztechnik

Herausgeber: Prof. Dr. G. Tränkle, Prof. Dr.-Ing. W. Heinrich

Cuvillier Verlag
Internationaler wissenschaftlicher Fachverlag

Innovationen mit Mikrowellen und Licht
Forschungsberichte aus dem Ferdinand-Braun-Institut, Leibniz-Institut für Höchstfrequenztechnik

Herausgeber: Prof. Dr. G. Tränkle, Prof. Dr.-Ing. W. Heinrich

Cuvillier Verlag
Internationaler wissenschaftlicher Fachverlag

www.ingramcontent.com/pod-product-compliance
Lightning Source LLC
Chambersburg PA
CBHW060449240326
41598CB00088B/4294